199 AGILE PRINCIPLES

FACTS • MYTHS • AND REFLECTIONS

Also by Brian Will

Agile Adoption and Transformation – Principles, Challenges, and Pitfalls

Agile Adoption and Transformation – Principles, Challenges, and Pitfalls
2020 Edition

199 AGILE PRINCIPLES

FACTS • MYTHS • AND REFLECTIONS

BRIAN WILL

FIRST EDITION
For permission requests, write to the publisher, addressed "Attention: Permissions Coordinator," at info@paroxys.com.

For U.S. trade bookstores and wholesalers order,
please contact the publisher via the website above.

Cover design by Ivica Jandrijevic
Interior layout and design by www.writingnights.org
Book preparation by Chad Robertson

ISBN: 978-1-7373374-0-9
Library of Congress Cataloging-in-publication Data:
Names: Will, Brian, author
Title: 199 Agile Principles / Brian Will
Description: Paroxys Press, Reston, Virginia, 2021
Identifiers: ISBN 978-1-7373374-0-9 (Perfect bound) |
978-1-7373374-1-6 (eBook) |
Subjects: | Non-Fiction | Agile | Computer software | Business process |
Classification: Pending
LC record pending
Printed in the United States of America on acid-free paper.

24 23 22 21 20 8 7 6 5 4 3 2 1

To Sayuri—for giving me all your unconditional love.
To Ian—for showing me all the possibilities of an open mind.
To mom, dad, and my brother—I miss you.

CONTENTS

Table of Figures

INTRODUCTION

I decided to write this book during the onset of the COVID-19 pandemic. Working from home and spending lots of time on social media, I had plenty of time available to think through my favorite problem space: Agility, Agile, and Digital Transformations/Adoptions.

Having published a book before, I was painfully aware that many people nowadays do not take the time to fully absorb four hundred and fifty-plus pages. So, when it came to format, I decided to take another approach.

Just like social media posts and online discussions, why not put together short but concise chapters that provide bite-sized information? Why not organize those bites into core areas that often prove difficult for newcomers and experienced Agilists to understand? Instead of making readers go through an entire book before they can get value out of it, why not organize it so that you can jump around and pick and choose the sections most relevant to you?

Basic Ideas – What are the underlying basic ideas that enable agile organizations to thrive? As many teachers and coaches will tell you, the basics are essential to get to the next level of expertise. Master the basics, and understand the intent of the basic ideas, and you are well on your way.

Myths – When it comes to agility there are many myths out there. And there are just as many charlatans that will hustle their wares, their buzzwords, and their acronyms. This chapter tries to address the most common misconceptions head on.

Facts – After we clarified some of the myths, we will refocus and talk about Facts. There are some immutable facts that any Agilist needs to be aware of when embarking on an agile adventure.

Principles – The Principles section talks about the core principles that make Agile and Scrum and many other agile frameworks work – if applied correctly.

Leadership – The Leadership section addresses some of the most important leadership-related challenges that will make an agile transformation work; or not. Leaders ultimately have responsibility for making their companies create value by making their agile transformation work. Without leadership and strong dedication at the top, you will not succeed.

Teams – Where leaders have responsibility for setting the direction, Teams are at the core of making Agile and Scrum work. Teams are at the forefront of delivering value every day.

Coaches – Many companies will retain agile coaches to get them started on their Agile or Digital Transformation. Having a coach can be tremendously useful for many reasons, but you also must understand some of the challenges you might face when selecting a coach.

Anti-Patterns – A list of common agile anti-patterns. This is mostly fun reading to illustrate what to avoid.

Best Practices – A list of best practices that have worked for many organizations. Best practices are a topic of heated discussion amongst industry practitioners, after all, who decides what makes up a best practice or not? Fair enough, there is no definite list, but there are common practices that are followed by many organizations in the IT or Software Development space.

Reflections – My personal reflections about Agility, Agile, and Digital Transformations/Adoptions.

I come from a software engineering background and have spent about half of my career in commercial-off-the-shelf software development and the other half in professional services/consulting. As such, my perspective about agility is swayed by my experience in technology, not in business. Although I have worked extensively with business executives, it was frequently as part of software development or technology refresh effort—with the focus on software solutions enabling new digital business processes.

My hope is that this little book will help all kinds of people to either be or become more agile—from beginner to experienced professional, across all functional disciplines.

Part I
BASIC IDEAS

Underlying **Agile** and **Scrum** are some basic ideas that drive almost everything and heavily influence various frameworks and methodologies. Understand the basic ideas, and you understand Agile and Scrum.

Some of these ideas are common sense, some are borrowed from other fields, some have been known for a long time but just have not made it into the mainstream.

Whenever you hear about some technique or approach and you are trying to understand why they are being used, look at these basic ideas to orient yourself.

Chapter 1
Optimized Communication

Why use Agile or Scrum at all? You hear that question all the time. The short answer is to get things done faster. That is the core of it.

People might justify it with all kinds of reasons, but executives will tell you, none of that matters. The only thing that counts is to get things done faster. Product cycles in all industries have accelerated; competitive pressures drive agile adoption.

The right side of the **Agile Manifesto** contains all items that were perceived to take up more time than they are worth—from a 2001 point of view.[1]

Small teams working together have better communication, resulting in faster solutions.

Showing working software, instead of showing a presentation, establishes facts.

Looking the customer in the eye and getting their agreement right then and there is more practical and more expedient than enshrining the agreement in legalese contract language that takes three months to agree on.

And being responsive to your customers creates long-lasting business relationships.

Agile and Scrum at their core are all about optimizing communication amongst all parties involved because faster communication delivers faster results.

Chapter 2
Small Batches of Work (Small User Stories)

How do you eat an elephant? One bite at a time!

Decomposing your work into small batches allows you to:

- Deliver faster.
- Estimate more easily.
- Absorb slips more easily.
- Provide better predictability because there is less variability—management loves predictability, as do customers!
- Train fellow engineers more easily—big enough to give newbies freedom to explore their own path, small enough not to endanger the job at hand.
- Create burndown charts showing progress.

That is why user stories should not be bigger than a maximum of two to three person-days of effort. Big is bad (call them Features or Epics), small is good.

Every successful company breaks down its humongous effort into bite-sized chunks! For example, Apple releases a set of new hardware and software EVERY YEAR. I promise you that this breaks down into a gazillion small batches of work/user stories/whatever you want to call it.

Production lines work like this. Small batches/small activities enable predictable production of results. Big batches/activities pose risk.

Chapter 3
Ownership and Personal Accountability

One task, one owner, one person accountable. Simple. Organizations that operate by committee are notoriously ineffective.

Such organizations also tend to come up with middle-of-the-road solutions—neither spectacularly outstanding nor horribly bad. Just "blah." Personal ownership and accountability create excellence.

This does not mean that you cannot have multiple people working on an issue, but you need to have one owner per task, otherwise, how do you know who to ask? Who is accountable? Who is coordinating?

Nothing creates ownership and accountability better than seeing your name on an issue, instead of the "Systems Team" or whatever other moniker you come up with.

Personal ownership and accountability are based on trust. Do not make somebody the owner of an issue, hold them accountable, and then proceed to micro-manage them. Ownership and accountability also assume a degree of autonomy.

Chapter 4
Small Teams Equal Better and Faster Communication

Organizing your work into small teams allows them to communicate better, faster, intuitively, and more efficiently. Why?

Extensive research has been done on optimal team size considerations and it is generally acknowledged to be somewhere between five to eight, with some teams functioning well with up to twelve team members. Some good examples of teams that need to make rapid decisions and communicate efficiently are sports teams and military units.

Most sports teams are organized around this size (up to 12):

- Soccer: 11
- Basketball: 5
- Baseball: 9
- American Football: 11
- Field Hockey: 11
- Ice Hockey: 6

Military squads are organized around this size (6 to 10) as well.

So why are small teams more efficient and communicate better than larger teams? The driving factor is the number of communication points and interactions, which exponentially increase with team size.

The bigger the team, the more interactions are necessary, the fewer fast and meaningful communication can happen. Things slow down.

Small, autonomous, teams enable fast, efficient, decision-making. Big teams are slow. That is why startups rule.

Chapter 5
Localized Decision Making

Decisions need to be made as close to the problem as possible.

Toyota introduced this concept with their Toyota Production System and Andon cords—everybody on the production line was able to stop the line if they saw a problem.

Amazon has adopted the principle of the Andon Cord for the products that it ships. Support agents at Amazon can "pull the cord" when they suspect a problem with their inventory (poor quality or malfunctions). This stops all sales and shipments before the issue spreads to more customers.

This kind of responsiveness is only possible with local decision making. If a company like Toyota or Amazon were to escalate problems up the chain of command for decisioning, it might take hours if not days to decide, resulting in faulty cars rolling off the line or broken products being shipped. Which in turn would impact operations due to rework, recalls, etc.

Local decision making enables the most efficient decision making that optimizes the entire process.

The bigger and more distributed a company is, the less efficient decision making becomes unless you enable and allow for localized autonomy.

CHAPTER 6
CONTINUOUS LEARNING

Continuous learning in an **Agile** and **Scrum** context usually means that the team follows the Deming/Shewhart Cycle (Plan-Do-Check-Act).[1]

You plan a little.

You execute/implement a little.

You check your assumptions to see if things worked.

You act according to your findings—either take corrective action or start over.

A similar concept was created in the U.S. Air Force by John Boyd, called the OODA Loop (Observe-Orient-Decide-Act).[2] It is specific to the combat operations process but can easily be applied within a business context as well.

According to Wikipedia, "The OODA loop has become an important concept in litigation, business, law enforcement, and military strategy. According to Boyd, decision-making occurs in a recurring cycle of Observe–Orient–Decide–Act. An entity (whether an individual or an organization) that can process this cycle quickly, observing and reacting to unfolding events more rapidly than an opponent, can thereby "get inside" the opponent's decision cycle and gain the advantage."

Essentially, by continuously learning and adapting, you are outpacing the competition—on a personal or company level.

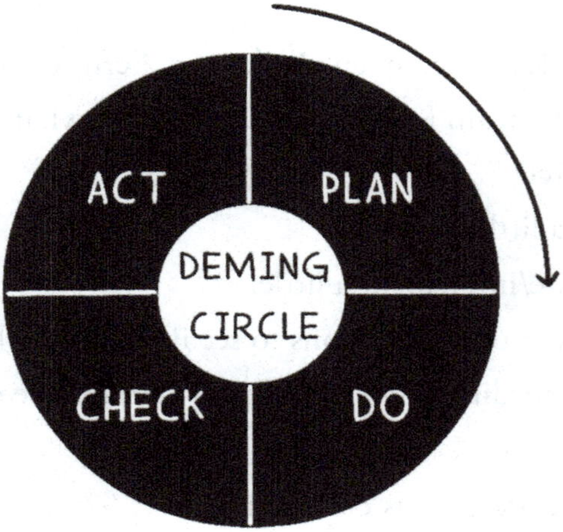

Figure 1 – Deming Cycle aka Shewart Cycle

Chapter 7
Value Streams and Wait Times

Entire books have been written about **value streams** and **value stream mapping.** The important parts to know in the context of software-driven enterprises are threefold:

1. There are two kinds of value streams. Operational value streams—value streams that provide business value to end-users and customers. And development value streams—value streams that provide automated business processes to be used by the operational value stream. To use the example of an insurance company, the operational value stream is the end-to-end process to provide insurance products to the customers, the development value stream is the process used to build the supporting IT systems or the actuarial research process to come up with new insurance products.
2. You need to know how long it takes to get through your value streams. This includes both process times and wait times.
3. Wait times can frequently be automated away or shortened by adjusting the work process.

If you desire to shorten time-to-market windows, you need to shorten value streams. DevOps processes and techniques shorten the development value stream and optimize value delivery.

Chapter 8
A New Way of Thinking, Planning, and Budgeting

In the old waterfall days, we tried to plan everything upfront, come up with a fixed requirements document (Scope), laid out an imaginary timeline on a project plan (Time), and then hired the folks to accomplish that (Resources). The Scope was fixed, and the Time and Resource estimates needed to implement the Scope resulted in a budget.

In **Agile**, we fix the Time and Resource and flex around the Scope. "You got 1 year, $3M, and 20 resources, show me what you can do!"

Because we always work off the most important items in the Backlog, we constantly deliver value. Backlogs are always prioritized, with top-level items explained in more detail which provides the most business value because these items are well understood. Every Backlog is arranged to show the most important items at the top and the least important items at the bottom.

You always work on the most important items—until your money runs out. That does not mean you cannot ask for more, but this way you are in a much better position because you delivered the most value before you did—which gives you credibility and the option to proceed or pivot.

This works on a $500K, $500M, or $5B program.

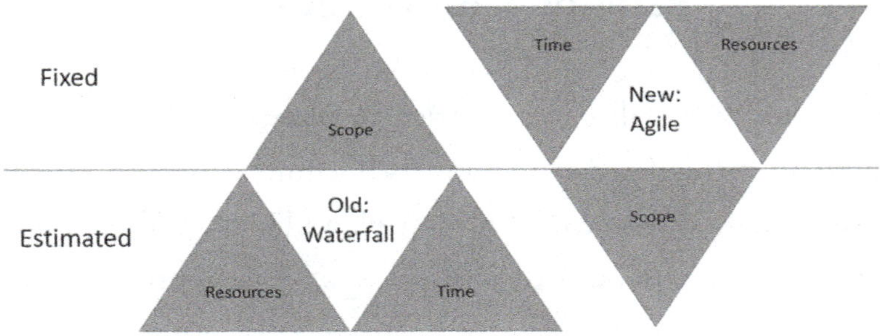

Figure 2 – Fixed versus Estimated Scope

Part II
AGILE MYTHS

Fantastical powers and unreasonable promises have been propagated by Agile and Scrum proponents. Many times, the depths of the arguments are paper thin, bordering more on wishful thinking than reality.

When it comes to agility there are many myths out there. And there are just as many charlatans that will hustle their wares, their buzzwords, and their acronyms.

All this is made worse by many applicable problem spaces that Agile and Scrum are used in—resulting in some statements that might well be true in one problem space but are utterly wrong in another. What might be true for a simple iPhone application can be dangerous and reckless for a medical device.

Following are some of the most common myths floating out there on the internet and in hallways.

Chapter 9
With Agile / Scrum, You Do Not Need to Plan Any Longer!

This is flat out wrong. You plan just as much as you did with other methodologies, but you plan differently—the biggest difference being that instead of big, upfront, planning, you now plan "as you go."

Agile planning turns traditional planning on its head. Instead of defining a plan based on upfront requirements (which may or may not hold true) that results in a cost estimate for a project, Agile fixes the cost and schedule and delivers against a flexible list of requirements that are prioritized in the backlog.

This prioritization implies that the requirements at the top are more valuable than the ones at the bottom and that the project is successful if the top ones are implemented. With every sprint, you re-prioritize, allowing the most valuable requirements to "bubble up."

Old style Waterfall methodologies try to plan everything in advance (define the requirements) to establish a planning and cost baseline upfront.

Agile, on the other hand, takes a fixed budget and schedule and asks: How many of the top priority requirements can we fit with the available money and time?

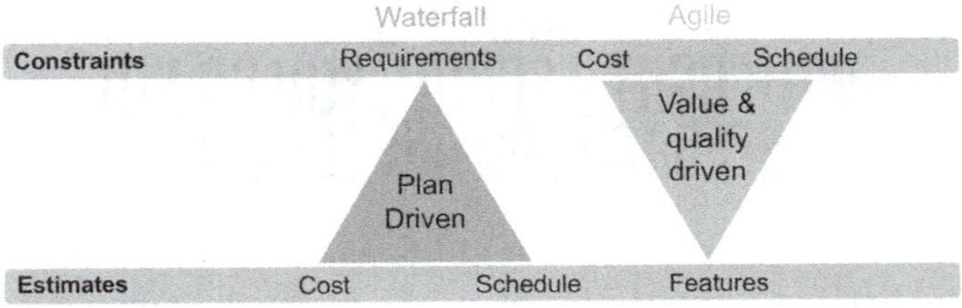

Figure 3 – Waterfall Schedule and Budget versus Agile Schedule and Budget

Chapter 10
Agility Guarantees Business Success

Although many Agile proponents nowadays argue as if one of the agile process frameworks will guarantee business success, it is simply not true.

There are lots of things that influence success: skills, tenacity, luck, management team, etc. Lots of books have been written about this subject, most of which do not mention following an agile process as a prerequisite.

Even startups do not insist agile processes are necessary for success. Now, most startups are agile, and some startups might follow an agile process, but none seem to highlight their agile processes as instrumental to their success. Also, remember that agile is now pretty much established across many industries, so it is not much of a competitive advantage any longer.

It is important to remember this.

So, what influences business success more than anything else? Market timing. I would submit that other factors like the management team or the agile process are important, but market timing seems to trump them all.

Yes, market timing. For an enlightening TED Talk about this, watch Bill Gross's: *The Single Biggest Reason Why Startups Succeed.*[1]

CHAPTER 11
AGILE / SCRUM WORKS ONLY FOR SOFTWARE!

Not true. Agile/Scrum is being used by companies in many different industries:

- Healthcare
- Subway Systems
- Rock Amplifiers
- Micro-miniature Medical Devices
- Construction
- Financial Services
- Banking
- Information Technology
- Human Resources
- Consulting Services
- Automotive
- Aerospace
- COTS
- Consumer Electronics
- Outsourcing
- Pharmaceuticals
- Oil/Gas

- Government
- General Business Management
- Etc., etc., etc.

Having said that, more and more industries are utilizing more and more software. The automotive industry is the prime example; cars essentially are driving computers now. Do not believe the "anti-hype" and buy into the "this will never work here" mentality. Agile/Scrum are here to stay.

CHAPTER 12
IT'S A JOURNEY...

It is one of the most common agile tropes you hear people say: "It's a journey..." Usually, it is in response to some challenge, problem, or non-agile behavior of the team or management. It seems to be the universal excuse that is used for not being agile. "It's a journey..."

Well, no it is not. At best, it is a side trip. Because becoming agile is not the end goal—making the business successful is!

Becoming more agile, driving the business to be more responsive, providing better value, or improving customer satisfaction are usually not goals that companies pursue in a vacuum. They are most likely based on competitive pressures—and those pressures make the problem of becoming more agile time-bound.

"It's a journey..." sounds like we are on a never-ending road of self improvement and self awareness. It completely ignores the competitive, financial, and business necessities that drive many companies to pursue agility in the first place.

An Agile Transformation is a means to an end, not an end in itself.

Chapter 13
Kanban Is Dramatically Different from Scrum

If you look at a **Kanban board** and a **Scrum board**, you will be hard-pressed to see a difference.

The biggest difference is that Scrum is time-bound (say, for example, to a two-week sprint), which means that everything on the Scrum board is supposed to get done and move from the leftmost column over to the rightmost "Done" column.

Kanban boards, on the other hand, flow; there is no time limit, and things get pulled into the board as needed and as capacity allows, then drop off the board when done.

They are different but related.

Scrum boards focus on achieving sprint increments and are ideal for coordinated synchronized events—like software releases, deliverables with hard deadlines.

Kanban boards focus on flow and work well for ongoing activities—like items coming off a production line, request queues.

Connoisseurs of Kanban or Scrum might argue about the details of this, but the differences are not that big.

Chapter 14
The Self-Organizing Team

The **Agile and Scrum community** is abuzz with fancy terms that are frequently misunderstood. One of them is the "Self-Organizing Team." Teams do not organize themselves. Team members are selected to be part of a team, and then the team organizes their own way of working.

The selection process of team members from existing staff is controversial. Should it happen at the team level (team decides) or at the organization level (management decides)? Are you assembling brand new teams, or replacing staff? New teams need to go through the traditional storming-forming-norming process and self-organize. This also needs to happen when replacing just one or two team members.

Management is still very much needed for selecting or hiring team members, as well as for disciplinary issues. Traditional Human Resources activities are always difficult, even in the best of companies.

Self-organizing teams organize their own work, but they do not form on their own. In this context, team member selection (a traditional Human Resources activity) is a critical activity. Choose the wrong team members, and you run the risk of permanently storming.

Chapter 15
Waterfall Was the Problem, Agile Is the Solution

Wrong! Waterfall was never the problem, the behaviors that a waterfall methodology encouraged were at issue.

There were always agile companies that followed a waterfall methodology but were not beholden to all the Waterfall problems - because folks had the right attitude.

Conversely, there are agile companies that follow one or the other agile methodology but cannot get things out the door - because folks have the wrong attitude.

Agility is a behavior, not a job title.

If you have good people, the process does not matter much. If you have the wrong people, the process will not help you.

Part III
AGILE FACTS

The opposite of myths are facts. Facts are based mostly on observable, verifiable, truths.

However, myths and facts sometimes go hand in hand.

For example, the first fact I list in this chapter is "Agile at scale is hard!" By all accounts, that is true; talk to anybody who ever endeavored to put a large company through an Agile or Digital Transformation and they will tell you "this is so hard!". What works well for one team might not work at all for 250 other teams.

But, for every realist that states how hard things are, you find others that capitalize on that truth and enthusiastically state "Agile at scale is not hard if you use our consulting services and methodology." Somebody took the unpleasant fact and turned it into an opportunity.

There is nothing wrong with that. Just be aware that those dynamics are at play. And remember that promises are easier made than kept.

Chapter 16
Agile at Scale Is Hard!

That is why most Enterprise Agile Transformation or Digital Transformation efforts fail!

Anybody telling you that an enterprise-wide Agile Transformation is easy simply has never done one before.

Beware of folks that talk a lot about transforming your enterprise but that have never done it before.

Most importantly, take responsibility for your own Agile or Digital Transformation—you cannot hire people to do this for you just as you cannot hire a dietician to lose weight for you.

Consultants can advise, train, mentor, and influence, but ultimately, you must transform yourself and own it.

Chapter 17
Do or Do Not. There Is No Try.

Go all-in with your Agile or Digital Transformation, or you will not succeed. Barely dipping your toes in it will not get you there.

Hernan Cortes burned his ships. Julius Caesar walled in his army at Alesia. Arnold Schwarzenegger hates plan B, as explained in the YouTube video ***"This is why Arnold Schwarzenegger hates plan B (motivation that will change your thinking forever)."***[1]

It is the same concept. What do we have to lose? Go all in! "Do or do not. There is no try!"

Who knew Yoda was talking about Agile back in 1980?

Chapter 18
Organizational Culture Sometimes Impedes Agile or Digital Transformation

Everybody talks about Agile. Every major consulting firm offers its Agile Transformation or Digital Transformation service.

Just like every football franchise owner hopes they will go to the playoffs; every CEO or executive thinks they can transform their organization.

However, the CEO or executive view of what their organization is capable of is based on what their middle management reports to them. Any assessment of cultural impediments will therefore pass through and be shaped by the system itself.

The information CEOs or executives receive about the organization, from the organization, will be inherently self-serving.

Agile and Digital Transformations are deep and pervasive, and understanding that it involves cultural change is not enough.

Reading about climbing Mount Everest does not make you a climber. Only experience does. Practice beats theory. Doing over reading.

Therein lies the problem—unless the CEO or sponsoring executive makes the transition herself, organizational change and Agile or Digital Transformation will remain elusive. They have to lead from the front.

Chapter 19
Quality Is Free

"The problem of quality management is not what people don't know about it. The problem is what they think they do know. In this regard, quality has much in common with sex. Everyone is for it. (Under certain conditions, of course.) Everyone feels they understand it. (Even though they wouldn't want to explain it.) Everyone thinks execution is only a matter of following natural inclinations. (After all, we do get along somehow.) And, of course, most people feel that all problems in these areas are caused by other people. (If only they would take time to do things right.)"

—Philip B. Crosby, *Quality is Free* (1979)[1]

Many of the tried and true quality systems thinking still applies, although some things have changed. Bottom line, if you are not producing quality product increments, you are not agile—you are most likely just adding technical debt.

You can find more information on this in my article ***Delivering Good Product Increments on LinkedIn***[2]

Chapter 20
Agility Is Relative

I have come across many folks who think that **Agility** is defined by X.

X usually is based on their own personal experience. Whatever your background and experience has exposed you to, determines your view of agility. I, myself, sometimes fall into that trap.

For example, after several years working in tech startups, sometimes literally in garages, joining more established outfits suddenly feels like wading through molasses. Whereas startups can make decisions sometimes in hours, more established companies may need months, quarters, or even years.

I have seen mobile development with daily sprints and monthly releases. I have seen large IT software integration programs run on 4-week sprint cycles and yearly releases (or even every two years).

Some deliver every week, some every quarter, some once a year. Sometimes your customers do not want more frequent releases.

There are big differences between short-lived mobile apps, development tools, banking software, IT integration projects, autonomous vehicle software, or mission-critical systems for the military.

It is all relative to your market. It is all relative to what your customers need.

I take it back: You might look a bit clumsy on land, but boy you're amazingly agile in the water!

Figure 4 – Agility is Relative

Chapter 21
There Are Few "Right" Technical Answers

Technologists come in a close second to religious fanatics when it comes to their preferred choice of cloud service, database, programming language, operating system, or whatever other technology to choose.

Fact is that unless you are picking and choosing brand new, cutting-edge, technologies, you usually have multiple good choices available. Just like in home improvement, where you have Lowe's, Home Depot, and ACE.

In cloud providers, you have AWS, Azure, and others.

In databases, you have Oracle, SQLServer, MySQL, MongoDB, and others.

In desktop operating systems, you have Windows, MacOS, and Linux variants.

In mobile phones, you have Android and iOS.

In programming languages, you have Java, Python, and many other choices.

Usually, you have at least two good choices, which is in line with the laws of market consolidation. Why is this important? Because you need to know that there are no "right" answers.

Remember, choices have consequences (think about licensing

costs for Oracle vs. open-source MySQL) and might paint you into a corner down the line. And sometimes a good choice might change over time.

Chapter 22
WORK STYLE INFLUENCES PROCESS

Work styles come in all kinds of variations. From the noisy, seemingly chaotic, environment of a trading floor to the quiet concentration of individual engineers working in offices. Some organizations are seeded with early risers, starting in the early hours of the day, some roll in leisurely around 10:00 AM. Some dress formal, some dress business casual, some casual, some barely dress at all.

How people work, their style of work, and even how they dress, highly influences the process you can follow. Shorts, T-shirts, and flip flops wearers usually are less process-oriented and communicate via text messages. Suits, ties, wearers of Wing-tip Oxfords most likely communicate via PowerPoint presentations and email attachments.

These are generalizations, but look around and see where your company fits in. Workstyle matters.

You cannot determine a process without considering the work style of your teams. And you cannot really mix and match them either—one process might not fit all.

Consider the work style of your teams or risk causing friction even before you get started.

Chapter 23
Management Is
an Artificial Construct

Although companies cannot operate without management, it is a completely artificial construct.

Many managers do not feel comfortable in their own skin, and many managers have attained their position not because of acknowledged managerial skills but because of a myriad of other reasons: prior title, degree, looks, connections, etc. These reasons usually are no indicator of managerial success. Remember the Peter Principle?

Understand that the way we organize companies today goes back to Max Weber (~1900), based on his review of the Prussian Army and the Catholic Church. According to the bureaucratic theory of Weber, bureaucracy is the basis for the systematic formation of any organization and is designed to ensure efficiency and economic effectiveness.[1] It is an ideal model for management and its administration to bring an organization's power structure into focus using hierarchical organization charts. We know now that this is wrong.

Agile/Scrum teams need to understand that. Although we all strive for recognition by a good manager, in the real world that frequently will not happen.

Chapter 24
Decisions Are Not Forever (Few of Them at Least)

Things keep changing and so decisions need to be revisited frequently. I have seen "Microsoft Only" shops suddenly start developing on AWS, Android developers branching out to iOS, and Oracle database zealots switching to MySQL or MongoDB. Most decisions are not forever. And some decisions will look pretty silly in retrospect.

If your company is truly market-driven, you need to let the market drive you. In the aforementioned example of Android developers suddenly embracing iOS as a supported platform, the decision was not based on their love for iOS but the fact at that time 80% of all app store revenues came out of the Apple app store. So, they simply made the right business decision based on market conditions.

Sometimes the driving factors might not be market-driven but driven by the desire to save costs or new technology capabilities. Whatever the reasons, decisions are not forever.

Chapter 25
There Is No Perfect Way to Measure Productivity in Software Development

When working with Agile teams I always encourage folks to look at their velocity—not because I believe it to be the best measure of productivity, but because it seems a good relative measure of team performance.

Why? Because in any complex endeavor, you need to separate the signal from the noise. For example, if you look at a soccer game, you might see a decisive win because one superstar striker lands four goals. But he was the recipient of many precision passes from his supporting team members. Does that make him more valuable than other teammates? No. What matters is that the team won.

In software development, it has been generally acknowledged that you might see productivity differences between team members that are measured in magnitudes. I have seen some engineers stare at complex problems for weeks, just to bang out a solution in 2 hours. And I have seen some engineers fruitlessly work on trivial problems forever.

Do not use simplistic 20th-century production line productivity measurements for complex 21st-century knowledge work.

Chapter 26
Software Keeps Changing

Many companies still believe that software and/or software-based infrastructure, such as servers or databases, do not change once a project is "done." Those are dying dinosaurs.

All the historical evidence points to the contrary. Software keeps changing (yearly/monthly/weekly releases), patches keep being applied (Patch Tuesday!), regulatory requirements force changes (CARES Act), and end-user behavior introduces changes that might not have been considered necessary (WFH, BYOD).

Fact is that software keeps changing at an ever-increasing pace and software is playing a more important role in most businesses.

The name of the game is to change constantly, to always have the latest and greatest infrastructure to run on—the latest software, the latest updates, the most current security patches. All while you iteratively and incrementally deliver your own product offerings.

Consequently, companies that are planning on being around in 20 years need to invest in cloud technology, Continuous Integration/Continuous Deployment pipelines, automation, and infrastructure resiliency. Or face extinction.

Chapter 27
Software Is Not an Asset—Starting Over Sometimes Is Best

Too many companies view software as an "**asset**," i.e. similar to an office building they acquired. Once implemented and operational, software is supposed to be amortized over years.

At a product level, it sometimes makes more sense to start over, to take the experience your teams gained from the first implementation effort and reimplement the software a better way.

It makes no sense to maintain and extend software that limits your growth potential or permanently boxes you into a corner.

At a company level, famous examples of this kind of "do-over" include Microsoft retiring MS-DOS in favor of Windows (while gaining experience implementing OS/2 for IBM), or Apple changing their chip architecture. These kinds of decisions are hard and require guts but often are strategically necessary.

The value of software is in the knowledge that teams gain from implementing the software, not in the lines of code themselves. Software is not a traditional asset. That is just the accounting view of it.

Chapter 28
If You Are Not Delivering Faster, Your Transformation Failed

Time to market windows, due to competitive pressures or new market conditions, are ever decreasing. Every company, in every industry, is in a race to deliver goods and services faster: Cars. Smartphones. Home mortgages. News. Credit cards. Groceries. Vaccines. Toys.

Disney missed out on big business because Baby Yoda was not available to buy when The Mandalorian aired. Agile fail.

The expectation nowadays is that goods and services are available instantaneously. Even if your product has not changed, customers demand better, faster, and cheaper service. Same day delivery and no-questions-asked return policies are standard.

Agile and Scrum are all about rapidly delivering business value. If your time to market windows have not significantly decreased after an Agile or Digital Transformation, it failed. You missed the boat. Essentially you replaced your old, slow, waterfall process with a new, slow, agile process.

Delivering business value faster is the only reason to go through a transformation. Nothing else matters much because nothing else will maintain your competitive edge for long.

Chapter 29
Good People Management Still Required

With all the agile process frameworks floating out there, you might think that process trumps people. Wrong!

Scrum, Lean, SAFe, DAD, etc. all describe agile processes, but mostly neglect (or just mention in passing) that the most important thing to consider are people.

Individuals and interactions over processes and tools. That is what the **Agile Manifesto** prominently mentioned on top of the value list.[1]

Somehow, we seem to have forgotten that good people management is still required and necessary to make agile work well:

- Be a leader.
- Take responsibility.
- Enable team success.
- Collaborate.
- State the purpose clearly.
- Manage emotions.
- Challenge assumptions and beliefs.
- Manage by trust.
- Thank employees for doing a good job.

- Listen to employees.
- Encourage new ideas.
- Create an open environment.
- Involve employees in decisions.
- Give employees a sense of ownership.
- Provide opportunities to grow and learn new skills.
- Celebrate success.
- Encourage cooperation.
- Do not micromanage.
- Be nice!
- Be positive!
- Have fun!

Remember, employees do not quit jobs, they quit managers—good people management is still required.

Chapter 30
Transformations Have a Beginning, Middle, and an End

A smart CIO once said, *"This transformation must have a clear beginning, middle, and an end. We want to get to a new steady-state—not transform forever."*

Beware of pretending that an Agile or Digital Transformation is an endless journey. Sooner or later, we all have to arrive.

Chapter 31
Cadence Drives Momentum and Delivery Performance

Ever wondered if your sprints/iterations should be four weeks, two weeks, one week, or one day?

Most discussions about short sprints center around how painful it is to have short sprints.

Because frequently, the perceived overhead (document, package, test, deploy) is too high for short sprints.

To adopt shorter sprints/iterations, you need to focus on automating those overhead items as much as possible.

Do not focus on how hard it is to do short sprints but ask what you have to automate to be able to shorten sprint cycles. Focus on momentum!

Automating overhead tasks will help you gain momentum and improve your delivery performance. The more you automate, the more momentum you gain.

Motion drives action, actions drive momentum. As you automate the entire Agile value stream, your momentum takes over and your Delivery Performance goes through the roof.

Lastly, do not have the slowest team set the cadence, but rather have the fastest team push their best automation practices to slower teams. That way even Agile at Scale programs will gain momentum.

Chapter 32
Most Agile Companies Do Not Talk About Agility

The goal for most companies is to be successful. They want to be more responsive to their customers, beat the competition, tackle changing market conditions, build value, and be profitable.

Most successful companies do not talk about "being agile" they just relentlessly improve upon their business. Look at Netflix, Mercedes-Benz, Google, Foxconn, Microsoft, Boeing, Facebook, Huawei, Apple, Bosch, Procter & Gamble, Amazon, and many others.

When embarking on a Digital or Agile Transformation, keep in mind that you will have a beginning, a middle, and an end. The goal is to transform and attain a new level of capability.

Do not become like the guy in the gym, who constantly talks about training routines and diets, but whose body has not changed in 20 years!

Transform, get it done, and be successful. If your company keeps talking about being Agile, you probably are not.

Chapter 33
Agility is a Behavior, Not a Job Title

Organizations, people, and politicians use the word "agility" as if it is a magic potion. Most do not understand what it means.

NASA is bureaucratic, slow, and fossilized. SpaceX is agile.

Agility is a behavior, not a job title.

Chapter 34
Software Does Not Age Gracefully

Old software has problems running on new Operating Systems. Old software does not run on many devices. Old software frequently is no longer supported or the support is very expensive. Old software is hard to maintain because nobody wants to work on it. Old software is prone to security issues. Old software frequently runs on old hardware which is hard to maintain and replace. Old software does not attract top talent.

Get rid of it!

New software has newer and better functionality. New software has fewer bugs. New software has fewer security vulnerabilities. New software usually works across many devices. New software frequently is cloud-hosted (which solves all kinds of other problems). New software is easier to maintain. New software attracts top talent.

I know that new is not always better. But in the case of software, new solves a whole set of issues that allows you to focus on your core business.

Chapter 35
Transformations Are Not Panaceas to All Business Problems

Your company history, how you have used technology, how you have engaged employees, how customers perceive you, and the markets you operate in, will strongly influence what you can accomplish in terms of an Agile and Digital Transformation.

Few companies are successful in trying to remake themselves operationally, and if they do, their performance often only improves marginally.

Be realistic about what an Agile and Digital Transformation can do for your existing business. Transformations are not a universal remedy to general business woes like changing market conditions or sales problems. Beware of consultants telling you otherwise.

Chapter 36
You Always Overestimate Capacity and Underestimate Problems

Bill Gates famously said that we have a tendency to overestimate what we can accomplish in the short run, and underestimate what we can accomplish in the long run.

Think about it. You hear stories like this all the time. I want to lose twenty pounds in four weeks! Wow, my neighbor lost a hundred and thirty in two years. Over/underestimation is all around us.

Agile and Scrum teams always struggle with their capacity. We will work forty hours a week so that we can get it done in four two-week sprints.

Not quite. Water cooler chatter–two hours. One-on-one with your boss–one hour. Meetings–eight hours. Bathroom breaks–two hours. Email–four hours. Online Chat–two hours.

Before you know it, you are down to twenty hours of dedicated, uninterrupted work time.

Top that off with unanticipated technical problems, the occasional sick day, or your five-year old having a fever, and all good capacity planning assumptions go out the door.

When planning your Sprints, Program Increments, or future Releases be conservative with your capacity assumptions or you are planning for failure.

Chapter 37
Scope Has Never Been Fixed

My waterfall friends always seem to get hung up on "Fixed Scope." They look at me and emphatically say things like "You do not understand, this is a fixed price, fixed scope project!" But the reality is that scope has never been fixed. In no project or program. Read my lips: Never!

We THINK that things are fixed because we try to express them in contractual terms to have recourse if things go off the rails. And what are the contractual solutions to changing scope? Change Orders. The contract language does not match what we are doing, so we will change the language and add a page, describing the new reality.

Even in the most simplistic problem areas, the scope is not fixed.

- You decide to go to an Italian restaurant and spend $50 on dinner with your wife. You end up at a Thai place and spend $80.
- You decide to buy a red Toyota Camry LE and spend $28K including all fees. You end up in a blue Camry Hybrid LE for $31K.

- Anybody who ever bought a house knows how flexible the scope is. Granite counters or a good school district?
- Do I have to take the trash out today? Maybe.

Flexible scope is all around us. Just acknowledge it.

- Agile design, flexible implementation!

Figure 5 – Flexible Scope, Flexible Implementation

Chapter 38
Pivoting from Old Technology Feels Uncomfortable but Keeps Jobs

Software, hardware, and technology, in general, keep changing at an accelerating pace. So much so, that people are resisting the change. Who wants to relearn all the time?

At a human level, this is understandable. Change is uncomfortable. At a company level, this resistance to change can be devastating to the viability of the company.

Look at retailers that for years have resisted the move to online sales and e-commerce. Or companies that did not optimize their supply chains both for speed and resiliency. Or car companies not focusing on electric cars.

Companies need to evaluate and assess their technology stack and infrastructure regularly and make hard decisions to pivot away from outdated technology processes. Adjusting their technology strategy accordingly is essential.

This kind of pivot can be hard because it might impact staff who need to be retrained, upskilled, or even replaced.

But those hard pivots are the decisions that ensure the long-term viability of companies. Which in turn provides jobs. That is the paradox.

Chapter 39
Agile Is Easy—People Are Hard!

Explaining and establishing an Agile and Scrum process is easy; dealing with personalities and team conflict is hard.

Chapter 40
Compromising on Agile and Scrum Processes Is Not Necessarily Bad

"Ideals may tell us something important about what we like to be. But compromises tell us who we are."—Avishai Margalit

We all start with the desire to create the perfect process, the perfect team, and the perfect product. In our minds, the perfect Agile or Scrum process will solve all problems.

That ideal is usually followed by us getting deflated by the realities of daily work life. As soon as we walk into a new or existing project environment, we start compromising. And that is OK. Making the right compromise is good.

The details do not matter much if we get the job done without compromising the Agile ideals.

There are no absolute truths about Agile and Scrum—because "Scrum itself is a simple framework for effective team collaboration on complex products." It is simple. It also is the basis for most Agile scaling frameworks.

Focus on compromising on the right things, so you live up to the Agile and Scrum spirit of the law, not the letter of the law.

However, do not compromise on the wrong things or you will undermine this very spirit. Find your True North.

Chapter 41
The Real World Has More Than Three Roles

Agile purists argue that you only need three **roles**: Product Owner, Scrum Master, and Team Member.

Depending on the agile framework, methodology, or toolkit you use, you might come across various (in some cases many) other roles.

If you are going through an Agile or Digital Transformation, you might have a long list of legacy roles that represent your current organization.

The reality is that you need to think through your problem space and come up with a set of roles that make sense for your organization.

Admittedly this is no easy task or something you can simply pull off the shelf. Many factors flow into this including industry, company size, history, regulatory issues, etc.

Do not fall into the trap of believing that there are a fixed number of roles in Agile/Scrum. Remember, there is a difference between a role and a job title. You can have nine team members with nine different job titles spread out across three or more roles. It all depends on your problem space.

Part IV

AGILE PRINCIPLES

Agile and **Scrum** are based on a set of principles that lay the foundation of the system. Without being aware of these principles, you run the risk of either failing or only achieving limited success.

Keep the following principles in mind to avoid common problems and pitfalls.

Chapter 42
Individuals and Interactions over Processes and Tools

Agile or Digital Transformation programs, especially for large companies, need to focus on hiring and placing the right people into critical transformation roles. Focus on people is critical.

Many companies are spending large amounts of money on Agile Transformation consultants, trainers, process documentation, and tools without adjusting the people side of the equation.

Although training existing staff in new Agile approaches, tools, and techniques is a must-do activity for any company undergoing an Agile or Digital Transformation, the more important focus area needs to be the hiring process.

I have seen companies that are completely out of balance when it comes to enabling Agile processes across the enterprise on the one hand but not hiring the right level of talent to make that new way of working successfully.

The best analogy I can think of is trying to run and recruit for an NFL team with a high school team recruiting approach. Hiring your future Agile leaders is critical to the long-term success of your transformation.

Chapter 43
Anticipate Organizational Gravity During Your Digital Transformation

Just like a fighter jet taking off, it is likely that organizational gravity (and people!) will pull really hard against changes resulting in reverting to traditional ways of working.

This is normal as most organizations operate under a presumption in favor of the status quo—anybody wanting to change an organization has the burden of proof to demonstrate that the new ways are more effective.

Change on a personal level is hard—think about how challenging it is to quit smoking, lose weight, or start exercising—at an organizational level it is infinitely harder!

Employees may observe old procedures and protocols while emulating new agile practices as well—this is the common anti-pattern of "doing agile" without "being agile."

Then there is personal pride, prejudice, and uncertainty–people might behave defensively when faced with the expectation of genuine change to their job, their influence, or the size of their office.

Anticipate this kind of organizational gravity pulling you back and counteract it or run the risk of your Digital Transformation failing.

Chapter 44
Do Not Ignore Human Resource Considerations

I have seen countless Digital Transformations where Human Resources (HR) considerations are completely ignored, especially around job descriptions, career paths, and career development.

While the IT/Engineering/Software Development organizations happily proceed to focus internally on becoming more agile, the HR function is ignored or kept out of the loop.

But there are many HR-related considerations that, if not taken care of, seriously impact the long-term viability of any Agile Transformation process. For example, conflicting job titles and roles and responsibilities.

Executives need to drive the HR considerations as soon as they commit to an Agile or Digital Transformation. This is a highly sensitive area that can derail the initiative unless things are communicated clearly across the board.

Chapter 45
Understanding "Plateauing" and "Zigzagging" Improvement Patterns

The good thing is that Agile/Scrum lends itself to empirical measurement. The bad thing is that many times companies expect linear improvements and do not understand the zigzag, course corrections, and plateauing (or flatlining) that a massive Digital or Agile Transformation goes through.

Improvements will most likely plateau (flatline) at some point, and look something like this:

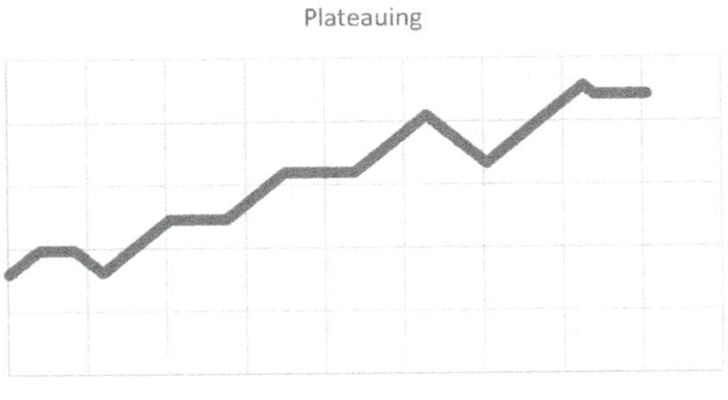

Or slide back and forth, and you will see measurable productivity loss, like this:

Figure 6 – Zigzagging and Plateauing Patterns

Frequently, executives (and middle managers who might have been critical to begin with) immediately panic when they hit the first bump in the road. "You see, this Agile thing will never work!" This is akin to calling off your diet as soon as you encounter the first meal that requires an adjusted palette. What did you expect? A walk on the beach?

It is critical to stay the course and work through these rough patches. Things will improve, and you will reach another high before you revert again. Continuous improvement over sustained periods of time does not happen in a linear fashion.

Chapter 46
Do Not Map Agile Into an Existing Process Framework

All established companies have existing methodologies and process frameworks. When embarking upon an **Agile or Digital Transformation**, it is tempting to overlap or map the new Agile/Scrum process into the existing methodology. Why not keep everybody happy?

For example, in the traditional waterfall model, you have the usual phases: Requirements, Design, Implementation, Verification, and Maintenance.

A common anti-pattern consists of mapping the new Agile/Scrum process onto existing process phases–you might hear something like "We are in sprint 12 of the Requirements Phase!" or "In this sprint, we'll do Requirements; the next sprint we'll focus on Design."

No, no, no! Do not do this! Bad. Very bad!

You will create a mess of a hybrid, with none of the benefits you are looking for. In the majority of the cases, this results not in the best of both worlds, but the worst.

Chapter 47
Do Not Change Agile / Scrum!

While undergoing an Agile or Digital Transformation, many companies will openly state that straight up **Agile/Scrum** just will not work and that Agile/Scrum needs to be modified to suit their special circumstances.

However, instead of improving on the overall delivery capabilities, creating this kind of hybrid solution oftentimes negates the benefits of Agile/Scrum. Avoid customization until you really understand the benefits. Why customize? What are the benefits?

Many times, process customization is a compromise to appease structural or political dynamics, instead of focusing on better value delivery. It always runs the risk of preventing the company from realizing the benefits that an Agile or Digital Transformation promises.

Sticking with standards reduces variation and enables learning from others. Customization puts you on an island by yourself.

This does not mean customization is bad, just that the company needs to understand why and what benefits are expected from it. By default, I recommend sticking with the standard.

Chapter 48
Understand Value Stream Mapping and Flow

Agile practice is really about understanding the basics of empirical process control in a complex and uncertain environment. Agile/Scrum is essentially a feedback loop that establishes empirical facts–done right, you know what the team's velocity is. Therefore, you can predict, within a margin of error, how much work can be accomplished in the future.

Those iterative and incremental techniques are geared toward experimentation. Agile practice helps an organization to learn about the business context in which it truly operates.

Value stream mapping allows executives to understand where there are critical bottlenecks, work-in-process limits, and flow that allows for a realistic assessment of what can be put through the system.

It never fails to amaze me to see a CEO or senior executive turn pale when you explain that 98% of a project's time spent is "wait time," not "process time." These are usually eye-opening discussions that help focus the executive ranks on the benefits of an Agile or Digital Transformation.

Chapter 49
Self-Organization <> Anarchy

Self-organization does not mean anarchy. Companies can still set standards, have guiding architectures, and change control processes but the teams need to be empowered to make the "on the ground" decisions.

Think about executives providing direction, but teams having the autonomy to implement based on their situational awareness.

Whenever you have extensive approval processes up and down the chain of command, with long periods of waiting for decisions, you emasculated your teams—and you lose productivity.

Fast decision making, by folks that know the problem space best (your teams!), enables fast movement, fast failing, fast learning, and fast value delivery.

Chapter 50
Empirical Process Control

I have been part of countless Agile or Digital Transformation efforts where the end goals just were not clear. "Do that Agile thing!" "Coach team X." Undertaking a massive Agile Transformation requires due diligence and serious consideration about how to measure success.

Empirical process control is typically lacking. A myriad of agile techniques will be introduced, with no clear target. Teams will be asked to deliver faster, but nobody will take the time to do value stream mapping to understand where time is spent.

Usually, teams end up doing agile by following all prescribed motions, but productivity does not improve. Empirical data is not gathered, analyzed, and discussed. So, true improvements remain elusive.

Executives must look for empirical evidence of improvement, not promissory notes, to be assured that an agile change initiative is indeed "on course."

Enterprise agility really means empirical mastery of innovation at scale. Senior executives must sponsor transformation across the enterprise and advocate the empirical process control that always underpins agile practice.

Chapter 51
PEOPLE OVER PROCESS

Do not spend your time, money, and energy on fancy process documentation!

Instead, focus your time, money, and energy on hiring the best and the brightest.

Of course, the size of your organization, the industry, and the regulatory environment many times determine how much of your process you need to document.

If you have the right people, the process will not matter much, and your people will be able to adapt as things change. And if you have the wrong people, the process will not save you and might even calcify your organizational capabilities. I have seen way too many organizations that are stuck in endless process discussions without creating value.

The process is not your product. Your people are providing capabilities that ultimately generate business value in your product. People over Process!

Chapter 52
Continuous Integration

Software architecture has moved towards more and more modularity and component-based systems. Cloud-based solutions have become ubiquitous. This new approach results in many independently running components and modules that need to be continuously integrated and tested.

Long integration cycles, and traditional branching and merging, have given way to trunk-based development. This means that on large, complex, systems all development teams are now "working in the same branch"–all the time.

Code changes are automatically validated every time code is run locally as well as when it is committed back to the trunk—daily or multiple times a day.

These kinds of new technical approaches allow for rapid deployments and rapid rollbacks. Rapid rollbacks in case of unforeseen problems enhance system stability and customer satisfaction.

Understanding continuous integration principles and goals is essential in making an Agile or Digital Transformation work–because otherwise you simply use new terminology without gaining the cycle time benefits you are looking for.

Chapter 53
CI/CD Is Enabled by Free Compute Cycles and Low Storage Costs

The reason that we can now accelerate software delivery cycles to the point of **Continuous Integration/Continuous Deployment (CI/CD)** is simple: Compute cycles and storage costs are (almost) free.

In the old days of computing dinosaurs (as recent as the 1990s), programmers had to carefully plan and think through the problem because compiling big projects (say 1M lines of code) took hours, sometimes days. Branching and merging code was a nightmare.

In the really old times, before even dinosaurs roamed the earth (say the 1970s), you had to encode things on cards, pass them to an operator, wait for your time slot, have the operator run things. Five days later, if you were lucky, you got a listing of errors. Rinse and repeat. Three months later, after your program worked, the requirements had changed.

How far we have come! Why? You can compile 1M lines of code on your laptop, which now probably has a 2 TB SSD. All the time. Over and over. Experiment. Get instant feedback.

If you are not compiling and running your code all the time, you are just letting your systems sit idle.

Chapter 54
CHOICES PROVIDE OPTIONALITY

Having choices is the essence of agility. Choices provide options.

Never staff to a single point of failure—always have at least two experienced engineers with relevant skill sets and a third one in training.

Make build vs. buy vs. rent decisions—do you need to buy servers and laptops? Can you use AWS or Azure? Buy laptops, lease laptops, or use Chrome books?

Do you need to have a big office building or can you settle on a small one with a part of your workforce working from home?

Do you have the in-house expertise to build a solution or do you have to hire? Bring in contractors? Partner with a service provider?

The more choices you have, the more likely you will have good outcomes.

Lastly, do not underestimate the lost opportunity costs. If you have five engineers and you suddenly must dedicate two of them to an effort you did not want to spend resources on, you did not just divert two engineers, you diverted four engineers.

Choices provide you the flexibility to deliver optimally. The fewer choices you have, the less agile you will be because you are forced down one specific path of action.

Chapter 55
You Must Understand the Real Organization

Agile and Digital Transformations suffer from blue-sky thinking, meaning not acknowledging the realities of the present organization that is supposed to undergo a transformation.

Designing a new agile organization on paper is simple but ignores the hard realities of how many things will have to change. Change the way people work. Change the way the company is organized. Change job titles and career paths. The list goes on. The larger the organization, the bigger the challenges.

Inevitably the existing organizational gravity will fight back against the proposed changes. So, to make an Agile or Digital Transformation successful, you must understand the real organization and the real political powers within a company to even have a chance of success.

If you do not, you inevitably get stuck with zombie transformations. People even have funny words for this: Wagile (waterfall-agile), Fragile (a riff on the difficulty of making Agile/Scrum work). The reality is that many organizations never make it to the future agile end-state they envisioned.

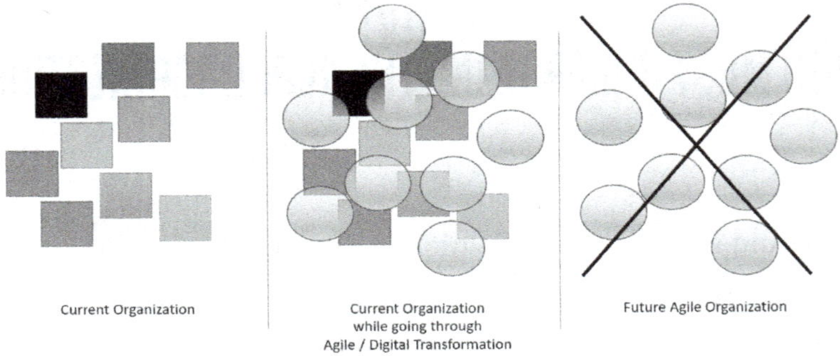

Figure 7 – Progression of Agile or Digital Transformation

Chapter 56
Ruthlessly Prioritize—The Magic of the Backlog

One of the key tools and approaches that are misunderstood by people new to Agile/Scrum is the **backlog**.

The backlog in its essence is simply a to-do list for your product or program. Traditionally the backlog consists of epics and user stories, but various Agile tools introduce other types of issues as well. Backlogs are frequently updated by adding, changing, or deleting user stories.

The overall product backlog is usually broken down according to planning/execution horizons–hence you will hear about the sprint backlog, or the release backlog, or the Program Increment backlog if you are working in a SAFe environment.

The magic of the backlog is that it is a prioritized stack. The most important and actionable items are on top and the least important items are towards the bottom of the stack.

By ruthlessly prioritizing your backlog, you constantly deliver the highest value items in your execution horizons. Lower value items will drop off the radar. Essentially you are flexing around scope, not around resources and time as we did in the traditional waterfall methodology.

Prioritization by business value makes the magic.

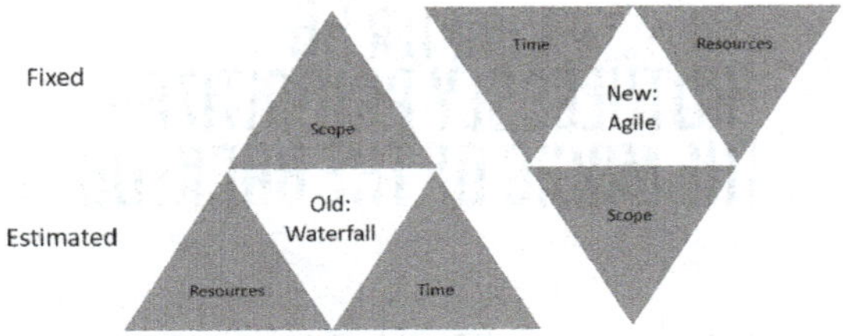

Figure 8 – Prioritization by Business Value

Chapter 57
Creative Conflict

High performing teams go through creative conflict all the time. When collaborating, high performing teams argue, counter-argue, disagree, some might even say they fight—but ultimately, they agree on a way forward.

This sort of creative conflict can cause some people to feel uncomfortable.

As a team, make sure that everybody understands that they can have their own point of view, that everybody can freely express their ideas, that everybody feels heard, but that ultimately you need to decide as a team and move on.

And individuals cannot pull the "I told you so" card if the decision turns out to be a bad one.

Creative conflict is a collaborative approach to get to the best possible solution–it is not personal. Embrace creative conflict as a way to drive towards the best solution.

Most of the time creative conflict will result in the right decision. If it does not, move on and fix it.

Chapter 58
Fix Problems, Not Symptoms

I once worked on a team that was very quick to fix things. Upon further analysis, we only ended up fixing symptoms, not the actual problems.

For example, when one of our Memcached objects got corrupted, one of our engineers would fix it right away. But he did not fix the cause, just the symptom.

Guess what? The problem came back with shocking regularity, along with 237 other problems, and before we knew it, we had one engineer spending all his time on "Production Support".

Fix problems completely. Kill them, so they never come back. That means you need to trace the symptom back to its root cause and fix the problem at its source. That is the only way you will be able to maintain good software.

Fixing symptoms should only be done to buy time to fix the problems.

Chapter 59
Fail Fast—Quick Decision-Making Trumps Analysis Paralysis

Failing fast is a concept based on small batch sizes (small increments of work) so that you never invest too much in guesswork, paper planning, or merely hope. You test your assumption quickly, either succeeding or "failing fast." The assumption is that when you fail fast, you can move on to the next theory, assumption, or approach to validate that in another experiment.

The concept of failing fast has been extended to DevOps and production systems as well. The idea is that due to complete DevOps automation of the Continuous Integration/Deployment pipeline you can roll out fixes quickly and see if the fix works—and if not, you failed fast, and due to that automation can roll back in short order.

Lastly, more and more companies are adopting the idea of failing fast to all kinds of business problems, where issues are discussed in small but competent teams, decisions are made in real time, and experiments are used to validate assumptions. The purpose in a business context is to avoid the proverbial analysis paralysis that large teams experience.

Chapter 60
Know Your Team Members and Help Them Grow

The Agile community is abuzz with discussions about I, T, or E-shaped staff. The thinking being that the more team members have deep expertise in several areas, with proven execution skills, and you let them freely explore and innovate, you will have created a high performing team.

Well, I could guarantee a regular Super Bowl win if only I had Thor on my team!

The problem with the I, T, or E-shaped staff discussion is that it seems wholly unrealistic for many companies:

- E-shaped specialists are hard to grow, harder to find, and almost impossible to keep.
- It assumes the availability of such specialists in the job market and your ability to hire them, including appropriate budgets.
- It assumes you can train somebody from an entry-level person to become I, then T, and ultimately E-shaped.
- It assumes employees want to become E-shaped specialists, and I have seen no evidence of that, especially among introverted technologists.

The best many companies can do is to know where your team members are in terms of career growth, help them grow, and create an environment that brings out the best in people. Do not wait for Thor to join the team!

I-Shaped Skills	T-Shaped Skills	E-Shaped Skills
Normal Employee	Rare "Jack of all Trades"	Thor
Specialist	Generalized Specialist	Future Specialist
Deep Expertise in one area	Deep Expertise in one area	Deep Expertise in few areas
Very few skills or experience in other areas	Broad skills across several areas	Experience across several areas Proven execution skills Constant exploration and innovation
Can create bottlenecks	Can remove bottlenecks	Can remove bottlenecks
Limited visibility to down stream impacts	Visibility and awareness to down stream impacts	Visibility and awareness across entire program
Impacts planning flexibility due to Single Point of Failure	Improves planning flexibility	Can create the illusion of planning flexibility due to unrealistic expectations

Figure 9 – List of I, T, and E-Shaped Characteristics

Chapter 61
Simplicity

The **Agile Manifesto** famously mentioned one of the principles: "Simplicity–the art of maximizing the amount of work not done–is essential."[1]

The KISS acronym–Keep It Simple Stupid–addressed this even before the Agile Manifesto was published. It can be applied to all kinds of problem areas: Architecture, coding, testing, personal organization, teamwork, etc. Simple wins over complex any time.

Every task you do not do, every presentation you do not have to spend hours editing, every meeting you can skip–basically maximizing the work not done–will provide you with more time you can focus on essential tasks.

Complexity is the enemy of great software. Simplicity is the great enabling principle.

Chapter 62
Read the Environment

Over my consulting career, I have seen hundreds of organizations, and I am always surprised to find that first impressions tell the story.

Fast-moving, dynamic organizations usually have a certain "hum" to them—that buzz of energy, laughter, hallway conversation, discussion, ad-hoc whiteboarding, all feeling somewhat frantic. Low energy organizations are, well, low energy. Folks seem withdrawn in their cubicles with no laughter and hushed discussions, while managers and executives are locked away behind closed doors in endless meetings.

When you are embarking on an Agile or Digital Transformation, read the environment. Take your first five days on the job to look around. What is wrong with this place? What works well? Do people seem happy and excited? Or submissive and depressed?

Why? This is the equivalent of a battlefield commander assessing the terrain. If you want to deploy big cannons, a swamp will not do—you need to move to more favorable terrain.

Transformations sometimes expose organizational challenges that have nothing to do with the actual transformation—and you have to solve for those first.

Part V
LEADERSHIP

Without visible, committed, leadership no Agile or Digital Transformation will be successful.

Leadership related challenges are common and are frequently quoted as the most critical failure points. Personally, I am not sure if it is mostly leadership challenges that are to blame for a transformation failure, but ultimately leaders have the responsibility for making their companies create value by making their agile transformation work.

Without leadership and strong dedication at the top, you will most likely not succeed, or at best, you will just be lucky.

Chapter 63
"Come Yourself or Send No One."

I am amazed when executives push back on the suggestion that they need to personally lead their organization's Agile or Digital Transformation.

Their pushback implies that "Agile is for those technical people." The quote above is based on a story (possibly fictional) that circulated about Dr. Deming during the late 1980s.

Supposedly the CEO of Nashua Corp initiated conversations with him via an introductory letter. It was signed by the CEO and suggested a time and place for the meeting.

A few days later a reply arrived by mail. It was addressed to the CEO and stated tersely, *"Come yourself or send no one."*

Dr. Deming knew that if senior leaders were not willing to lead an initiative from the start, then it would be stillborn.

Most organizations, especially large ones, are hierarchical so it is natural for executives to want to delegate tasks.

Unfortunately, an Agile or Digital Transformation cannot be delegated. Everyone must be involved in enterprise transformation, and the responsibility for managing its success cannot be passed on to someone else—the CEO and her most senior leadership team are on the hook.

Chapter 64
Lack of Executive Change

Executives cannot delegate their Digital or Agile Transformation effort—they must be part of it to lead it. That sounds great in the abstract, but what does it mean for executives in practice? Some examples might illustrate.

Executives at all levels need to:

- Flatten the organization.
- Change weekly status meetings to daily standups.
- Be accessible.
- Establish direct lines of communication with people doing the work.
- Provide and be open to receiving feedback freely.
- Go on Gemba walks.
- Create a sense of urgency.
- Focus on delivering value, not process.
- Lead by example.
- Stop using PowerPoint! When you present, you are talking at people—you need to talk to people!

The reluctance of executives to change themselves is a big challenge for any Digital or Agile Transformation effort.

CEOs beware! If you and your direct reports do not change, the organization will not change. You can read more about how PowerPoint is impacting organizations in the *Small Wars Journal* article "How PowerPoint Stifles Understanding, Creativity, and Innovation Within Your Organization"[1] or *The Guardian* article "How PowerPoint is Killing Critical Thought."[2]

Chapter 65
The Need to Go All-In

Hernan Cortes burned his ships. Julius Caesar walled in his army during the Siege of Alesia. The allies landed in Normandy on D-Day. Going all in is a prerequisite for success.

If you want your Digital Transformation to succeed, you need to go all-in—there is no dipping your toe and trying it. Or, as Yoda said, "Do or do not, there is no try."

Without full out, strong, unwavering, massive action and agile sponsorship at the executive level, organizational gravity will pull the organization right back into the status quo.

You know this instinctively. How many of us have tried to learn a foreign language one hour at a time? Did not work, right? But, if you move to Italy for a year, without a safety net, you will learn the language.

Going all in and taking massive action will yield results.

Dipping your toe, sort of, kind of, will not.

Chapter 66
Creating a Sense of Urgency

If you have not heard your CEO talk with urgency and specificity about the Agile or Digital Transformation underway, then, well, it probably is not.

A failure to communicate this imperative and set expectations across the enterprise, accordingly, will stop critical momentum from building up. The change will "fizzle out."

Executives must take care to ensure that Agile or Digital Transformation is given top priority, communicated effectively, and driven hard toward clear, measurable, goals.

After all, the end game is to be agile, and deliver real, tangible business value and improvements—not spend five years trying to do that Agile thing.

Chapter 67
Intermixing Program Efforts Will Risk Your Digital Transformation Success

It is not uncommon for companies to try to piggyback their Digital or Agile Transformation as part of another program.

For example, a company trying to leave old legacy systems behind by employing a lift and shift initiative might be tempted to go through an Agile Transformation at the same time.

This is not very likely to succeed. You do not modernize your military while fighting a war. You cannot upgrade your medical equipment while performing an operation. This is the organizational equivalent of multi-tasking, and we know that does not work.

The focus of the Agile Transformation will be diluted by the other program's challenges, and if the overall technical exercise experiences problems, the agile focus will take a backseat to "just getting it done."

Agile change is deep, pervasive, and must be expected to transcend all existing work and associated programs.

Sponsorship for agile practices must reach beyond one program manager or a set of senior managers occupied with other deep technical challenges—it can really only come from the very top.

Chapter 68
Executives Need to Reconsider Reports and Metrics

Executives often depend upon reports to give them an idea of what is going on and the decisions they ought to make.

Reports, charts, and statistics provide them with essential information. Executives interpret these sources using the indicators and measurements they think best reveal the underlying truth. Their views of the organization, the challenges, the projects, and the people are shaped by these reports and metrics.

Not adapting old reports, charts, and metrics to a new Agile environment sets you up for a lot of pain.

However, executives who look at standard Agile burnup and burndown metrics, familiarize themselves with the standard Agile information radiators, and focus on innovation accounting and the elicitation of actionable rather than vanity metrics achieve multiple wins at the same time.

They make reporting more effective. They avoid transcription issues that might occur if old reports and metrics stay in place, and they send a clear message to the Agile teams that the executive ranks are serious about the Agile or Digital Transformation.

Chapter 69
Establishing a Product Focus

Most large organizations act like they are in the business of running projects. But they are not. Most business value is sustained through ongoing product efforts.

Projects are by definition temporary endeavors. They have a beginning, a middle, and an end. Much effort is expended in setting them up, monitoring their supposed progress in the absence of regular useful deliverables, and tearing them down afterward.

Projects do not sustain the flow of value, nor do they retain the institutional knowledge and team skills that allow value to be optimized.

Business agility rethinks the enterprise not in terms of projects, but rather in terms of the genuine products that allow value to be released and empirical lessons to be learned.

Establishing a product-focused organization is undoubtedly a step forward to understanding value streams of any type, and it is a significant improvement compared to the traditional project-execution mode.

Senior executives should therefore take care to ensure that products rather than "projects" are being sustained and are clearly owned and represented at all levels.

"...AND THIS IS OUR AGILE PROJECT!"

Figure 10 – Product Focus vs. Project Focus

Chapter 70
Striking the Right Balance Between Alignment and Autonomy

The goal for any Agile or Digital Transformation is to create alignment of company goals (the top) with autonomy at the team level (the bottom).

Teams need autonomy. That is what drives them to come to work and exceed expectations. At the same time, work must be aligned with business goals. No alignment and no autonomy equal no results—or very poor results.

Too much control results in micromanagement, so everything slows to a crawl. Exert too little control and chaos ensues.

Management is responsible for the rules of the road. This includes clarifying the roles, teams, cadence, and terminology.

Teams have autonomy in terms of how they go about doing the work in both planning and practice. Within the overall framework, each can take a different approach. The specific engineering practices are up to the teams. For instance, whether the team decides to do pair programming is up to them.

The role of management is like establishing the rules of the road. The role of the team is like choosing the route to get to the destination.

Chapter 71
Letting Go of Control Is Hard—Focus on the Important

When I was a young manager, I can now easily see that I was a control freak. Rules, standardization, compliance, striving for excellence even in the most mundane of subject areas, worrying about every detail.

Then somebody promoted me to manage people. Oh my...

Everything my direct reports produced seemed not good enough. Why did she choose that non-standard color scheme in the Lotus 1-2-3 bar charts [If you do not know what Lotus 1-2-3 is, Google it!]? They said they would be done on Thursday and it slipped to Friday morning! Why was that paragraph so wordy?

I must have come across as stressed-out because one of my direct reports looked me straight in the eye and said, "Just let me deal with it, you do not need to worry about the details." So, I stepped back and discovered that the color of bar charts did not really matter, that there was not a substantial difference between COB Thursday or Friday morning, and that wordiness did not matter as long as the meaning was clear. Letting go of my insecurities made my direct reports happier, helped them grow, and ultimately produced better results. Focus on what is important, which is enabling your folks, not controlling them. Focus on the important.

Part VI
AGILE TEAMS

Where leaders have responsibility for setting the direction, Agile teams are at the core of making Agile and Scrum work. Teams are at the forefront of delivering value every day. Leadership defines the strategic "What?" Teams are responsible for "How?" To use an analogy, leadership sets the objective, say to drive from Los Angeles to New York. The team is responsible for the day-to-day driving decisions—where to turn, where to stop, when to rest, when to refuel, etc.

Just like in real life, if your leadership keeps changing their mind and redefining the objective to change from New York to Miami and then to San Francisco, your team will be bee-lining all over the map. Leadership is challenged.

Conversely, if the team is randomly changing the driving direction even though the objective has not changed, the team is challenged.

Chapter 72
The Team Is the Product!

Old way of thinking: We have a product and can hire anybody to make it great and sustain it. The product is an asset with a dollar value, supported by fungible staff. You can sell the product and do not have to worry about who works on it.

New way of thinking: Time to market windows are ever shortening. In software, release cycles have gone from years, to quarters, to months, to continuous updates. The asset is the team that is capable of delivering products. Good teams have a longer lifetime of generating value than the product itself.

Some companies have a hard time wrapping their mind around this, especially in a time of international labor arbitrage and offshoring/outsourcing.

But this is true for many "products": The US Marine Corps, Google, FC Barcelona, the New England Patriots—those great teams make their respective great product, not the other way around.

The team **is** the product!

Chapter 73
Hiring Good People Is Hard

If you genuinely want to become an agile company, hire the right people. With the right people, you will be able to adapt to any changes the business needs. With the wrong set of people, no process or framework in the world will help you.

- "Hire people who are better than you are, then leave them to get on with it. Look for people who will aim for the remarkable, who will not settle for the routine." —David Ogilvy
- "The best executive is the one who has sense enough to pick good men to do what he wants done, and self-restraint to keep from meddling with them while they do it." —Theodore Roosevelt
- "Hiring people is an art, not a science, and resumes can't tell you whether someone will fit into a company's culture. When you realize you've made a mistake, you need to cut your losses and move on." —Howard Schultz
- "Surround yourself with the best people you can find, delegate authority, and don't interfere as long as the policy you've decided upon is being carried out." —Ronald Reagan

Chapter 74
Deal with It or Live with It

Sometimes decisions we do not agree with are made. At that time, we all have a choice to make: Deal with it or live with it.

I have seen people and sometimes entire teams grumble for months, sometimes even years, about this or that decision that they "did not agree with."

As professionals we have a choice, if you do not like the decision, rise and make a credible argument for your alternative—deal with it. Or live with it, even if you must hold your nose.

There is no third option here, grumbling about things will not make you happier, will not help your team members, and it surely will not help the company.

Chapter 75
Listen for the Unspoken

You walk into a retrospective meeting and nobody seems to be willing to speak up. You attend a daily Scrum meeting, and it is always the same two or three team members that are talking, the rest of the team just listening. During the sprint demo, if one of the stakeholders is present, everybody seems to clam up.

Listen for the unspoken. What does it mean? What does it tell you?

Teamwork is essential nowadays, but frequently teams barely work together. There could be many things at play that prevent teams from openly communicating:

- Company culture.
- People got punished previously for speaking out.
- Stakeholders are the ones that will approve raises or bonuses.
- Cultural backgrounds might make it hard for young engineers to speak up against their "elders."
- Some people might be A-type dominant, not thinking twice that they always take the floor.
- Employees have not digested the change from a command-and-control-like environment to Agile and Scrum.

- Etc.

Listen for the unspoken. If you listen, you will quickly find out what the challenges are.

Chapter 76
Soft Skills Outweigh Hard Technical Skills

Building and maintaining an effective development team is not a matter of interviewing team members for their particular technical skill sets such as Java, C#, Oracle, or MySQL—what matters most is that team members are:

- Personable.
- Experienced.
- Reliable.
- Communicative.
- Cooperative.
- Freely share.
- Solve problems.
- Committed.
- Respectful.
- Supportive.
- Fun to be around.
- Have good aptitude.
- Want to be accountable.
- Bring a great attitude to the table.

In short, team members need the right soft skills to be effective. Hard technical skills can be taught.

Chapter 77
How to Interview Team Members

The most common problem with interviewing team members is that most companies put candidates through a day of thirty to forty-five minute, one-on-one interviews with a variety of interviewers. It is followed by some kind of assessment to be filled out by each interviewer, based upon which a decision is made. All this is generally inefficient, error-prone, and a cause of frustration.

So, what to do? Group interviews! Instead of eight thirty-minute individual sessions, organize one ninety-minute session with four to eight interviewers that interview the candidate interactively. Do role-playing. Have the candidate whiteboard things. Have a discussion. See how the candidate reacts under pressure. Let the candidate have a chance to see if he fits in and feels comfortable with the future group of peers.

Technical qualifications can often be assessed within the first part of the interview process. Personality and fit in the second part. Group interviews are most effective because all participants see the candidate in the same circumstances and hear the same answers.

Forget about three-letter certifications. Focus on attitude, aptitude, and accountability.

Chapter 78
Maintaining Good Teams

Maintaining good, effective, teams over the long run is a tricky process, as you need to balance several different demands.

- On the one hand, you want to maintain team stability—you want as few people as possible to leave, move to another department, or take some other responsibility outside of the development team
- On the other hand, you do want to create some kind of career progression, as well as lateral mobility so development team members do not feel "stuck."
- Then you have to deal with competitive pressures, local market conditions, and budget realities

The truth is that you will need to engage in a constant rebalancing act to maintain an effective development team. This rebalancing involves understanding a clear picture of all technical challenges represented by your projects, contrasted by the soft and hard skills required to deliver against those challenges.

This rebalancing might feel like you are never done analyzing, interviewing, and hiring—because you never really are.

Chapter 79
It's the Manager

If you are care about establishing effective Agile teams, check out this book from Gallup, *It's the Manager: Moving From Boss to Coach* by Jim Clifton and Jim Harter.[1] It is full of valuable insights and lots of data.

Chapter 80
Product Owners—What Makes a Good One?

Product Owners need business acumen—that is a given. Product Owners need to be domain experts or at least understand enough of the domain to pull information effectively from others. If the Product Owner does not understand the business, he really cannot own the product!

Product Owners need to understand the basic Agile/Scrum process they are operating within—that is a given also. Although many organizations make detailed knowledge of Agile/Scrum a must-have requirement when hiring Product Owners, it is fair to say that anybody interested can acquire Agile/Scrum process knowledge in two days of study or by attending a CSM/CPO class.

Product Owners need excellent people and communication skills—decide, communicate, facilitate, arbitrate—these are at the core of what a Product Owner needs to do.

Chapter 81
How to Hire Product Owners

I am constantly blown away by the concept of hiring a Product Owner on demand. Or that idea of a contract Product Owner.

How do you hire a Product Owner? The short answer is, you do not hire Product Owners—you grow them!

Above all, Product Owners need business acumen—for their specific business. And you do not get that by hiring somebody brand new. You need to grow that expertise in-house. Put your best and brightest into this job, you cannot just buy success!

Now, can you hire somebody talented and get them up to speed so they become a solid Product Owner? Yes, but expect an extended learning curve. Also, understand that this will be a high-risk endeavor. Just like other new hires, some work out, some do not, some might seek better opportunities after you hire them. Make sure to understand the trade-offs.

Chapter 82
Scrum Masters—What Makes a Good One?

Scrum Masters require a complex mix of soft skills, experience, and personality traits that are hard to nail down and even harder to hire for.

Scrum Masters need to be able to challenge others and be comfortable being challenged by others; their ability to ask for help, especially from higher-ups in order to remove obstacles, is useful but often a difficult task requiring political finesse.

Agile/Scrum emphasizes delivering customer value over extensive documentation. That is reflected in the Scrum Master's emphasis on engineering practices and focus on working software.

Agile/Scrum promotes open communication and active contributions from team members and the Product Owner throughout the project. An effective Scrum Master accomplishes this by encouraging and facilitating ongoing verbal collaboration.

Scrum Masters must be able to hold to their process convictions (the Agile/Scrum process), their own skills, and their team's ability to deliver. They are supposed to be the steady rock amongst the waves of uncertainties that development activities sometimes pose. That is why good ones are hard to find.

Chapter 83
How to Hire Scrum Masters

Anybody can pass a Scrum Master certification, but how effective will your Scrum Master be? Here is a subset of questions that need to be answered during the interview process:

1. Right personality?
2. Right technical skill sets?
3. Knows Agile/Scrum?
4. Can collaboratively influence the team?
5. Can influence stakeholders?
6. Can remove impediments?
7. Has the right soft skills to work collaboratively?
8. Has enough of a hard edge to not appear to be a pushover?
9. Has specific technology and business skills relevant to the effort under development?
10. Can draw on prior professional experience?
11. Has the ability to focus the team on the tasks at hand (sprint focus) while considering the big picture (release focus)?
12. What impediments were removed for the past teams?
13. What are at least two examples of the "servant" role that

were filled in the past?

14. What are at least two examples of the "leader" role that were filled in the past?
15. What process improvement suggestions failed? Why?
16. Two good examples of the candidate influencing stakeholders.
17. Can deal with criticism?
18. Examples of candidate coaching a prior team.
19. Exhibits intellectual curiosity?

Chapter 84
The Smarty-Pants Product Owner

The joys of working with people involve coming across all kinds of personalities.

The smarty-pants Product Owner is frequently characterized by high intellect but simply does not want to get stakeholders or subject matter experts involved in the refinement process. Basically, she believes that she knows it all.

A variation on this personality type is the overly dominant Product Owner. Again, oftentimes characterized by above-average smarts, she proceeds to create user stories by providing not just the "why" but also the "how" and the "what."

Remember, Product Owners are supposed to detail "why"; teams are supposed to answer "how"; and the collaboration between both parties results in what scope is necessary.

Product Owner: As a user, I need persistent storage, so that I do not lose data. – Why?

Team: We are going to store the data in a relational database. – How?

Collaboration between the Product Owner and the team is a discussion about what scope is necessary to achieve success. Will a small, free, SQLite database do? Or do we need an expensive, enterprise-level Oracle database with replicated instances in twelve time zones? – What?

Chapter 85
The Submissive Team

When submissive teams blindly follow whatever the Product Owner demands, two key problems arise:

1. How does the team know that they are delivering high value to the business? Challenging the Product Owner whether her choice of issues is the best use of the Development Team's time is critical to making sure you really hit high business value items and not just vanity activities.
2. The Development Team asking critical questions is the manifestation of cross-training. I have seen many teams become experts in the very areas they knew nothing about simply by asking "Why?" Before you know it, everybody builds up domain expertise!

From a team dynamic perspective, this can be caused by:

- Dominant, type A personality of the Product Owner combined with a very timid team.
- The team is not engaged—personality conflicts, career concerns, etc.

- Structural issues such as the team directly reporting to the Product Owner, meaning she has managerial authority over the team members

Whatever the reasons, if you see a submissive team, try to figure out what is going on because you are not getting the most value out of your employees.

Part VII

AGILE COACHES

Many companies will retain **Agile Coaches** to get them started on their Agile or Digital Transformation. Having a coach can be tremendously useful for many reasons, but you also must understand some of the challenges you might face when selecting a coach.

Coaches come in all kinds of varieties. Some are technical. Some are not. Many specialize in one or the other Agile or Scrum methodology. Many of them have countless certifications.

There really is no clear definition of what background a coach should have, although there are a growing number of "certification authorities" that claim to have the expertise to train and certify coaches.

On top of all this, many top-flight consulting firms offer their own Agile or Scrum methodologies, selling engagements to initiate and guide entire Agile or Digital Transformations—including coaching staff.

This section highlights some of the major challenges and shortcomings you might find in various coaches.

Chapter 86
When Your Coach Has No Domain Expertise

I was talking with colleagues recently and immediately had a sense of déjà vu. The topic was "Should a coach have domain expertise?"

I strongly believe that you need domain expertise. I am a software engineer by training. I managed software development teams of all kinds of sizes and technologies before I started coaching, and feel that this domain expertise makes me a better coach—for software development or IT projects.

Whenever I dabbled in business agility and hardware before, I felt like a fish out of water.

I always felt strongly that if you coach a soccer team, it helps if you played soccer before. My son suffered through one soccer season where the coach was a football coach—that did not work too well.

If you learn to dance, you would expect your coach to have danced before. If you want to build a custom chip production line, you would hope that your coach understands the hardware design process.

Beware of coaches that have no relevant domain expertise. At best, you are paying for their learning curve; at worst, you get bad advice.

Chapter 87
The Coach Is Not Meeting the Client Where They Are

Agile Coaches have a responsibility to meet their clients where they are—not where they wish them to be.

For example, if your coach has extensive experience and a high level of sophistication in terms of DevOps processes (i.e. automated unit testing, BDD, CI/CD automation, automated regression testing, on-demand deployments, etc.) and the client has no DevOps experience, the coach needs to coach at the appropriate level. Do not run NFL level drills for the pee wee football team!

The corollary of this is that clients have a responsibility to hire the right Agile coaches.

For example, if you expect a highly technical coach to help your teams automate the CI/CD pipelines, hire someone with a technical background and experience. A coach with no software development experience will provide little value.

Chapter 88
Coach Uses Buzzwords Over Explanations

Agile Coaches, including myself, tend to be full of themselves. After all, that is the reason we became coaches! We think we know best.

I cannot help but think about the last physical I got and the explanation that the doctor gave me. "Let us review your blood tests. Your CBC, BMP, and LFTs were basically negative. Your slightly elevated LDL, and since your BMI is 28, you should watch your diet and exercise more to prevent metabolic syndrome. And keep taking an occasional NSAID with food for your knee pain."

I am sorry, what?

Agile Coaches do the same thing. "In order to shorten the delivery cycles, we need to focus on making User Stories smaller. This will shift testing to the left and show us the burn down during sprint execution. It will also establish team velocity. Running extensive unit tests using JUnit and CI/CD processes using Jenkins will enable faster feedback loops. Automated build scripts can deploy Docker images that will run our BDD scenarios using Cucumber. Our team already has the Gherkin scripts."

Business stakeholders are not impressed by this. Speak plainly.

Chapter 89
Coach Is Too Touchy Feely

Some **Agile Coaches** focus 100% on people—their thinking usually goes like this: People are the most important thing in a company, if I focus on people, and make sure to bring out the best in them, the rest will fall in place.

Coaches like this are usually what I call "cuddly." They talk a lot about feelings, family, fun, and working together as a team to get things done.

Do not get me wrong, all of these are good things. Having staff members feel good makes a difference. However, that by itself is not enough. I have seen plenty of teams that felt good, appreciated, and respected, but did not deliver value. Guess what happens to teams like that?

I prefer the sports analogy of coaching. The coach is there to enable the team, help them grow, bring out the best in them, and hopefully get them to the playoffs. But coaching also involves hard discussions about teams not performing and recommending players be released or traded to another club.

Ultimately coaching is about creating business value and coaches are responsible to the business. Just making people feel good is not enough.

Chapter 90
Coach Believes Certifications Equal Knowledge

Many **Agile Coaches** confidently present themselves, stating: "I have an ABC certification; I know what I am doing."

Not necessarily. Certifications show you have rudimentary knowledge; they do not show what you actually can do in the real world, i.e. if you can apply the knowledge.

As Mike Tyson said, "Everyone has a plan 'til they get punched in the mouth."

Getting a certification is the equivalent to having a plan Congratulations! But that does not qualify you to successfully step into a real Agile project and be a successful coach. Only real experience can do that.

Chapter 91
Coach is Training Instead of Coaching

Some coaches are expert trainers. But that is rare. I have found that people that are good at coaching are not necessarily good at training.

Good trainers do not necessarily make good coaches. As a matter of fact, some of the best trainers I have come across were horrible coaches.

Good trainers are excellent at managing the classroom, the materials, the pacing, the exercises, keeping the class on track, throwing in the right jokes at the right time, while finishing on time. They cover the materials in a way so that most attendees comprehend them in the allotted time.

Good coaches are more random. They mostly listen, watch for signals from team members, observe behaviors, and suggest corrections to bad behaviors in a way that does not put folks on the spot. They guide instead of direct. They will not talk too much and will not take center stage. They freely share praise without becoming overbearing or patronizing. They effectively communicate at all levels of the organization.

Coaches can do training, and trainers can do coaching. But the talents exhibited determine what they are.

Chapter 92
When the Coach Is Too Tools Centric

A fool with a tool is still a fool—or so the old saying goes.

Some coaches are great with sticky notes and index cards but struggle with online tools like Jira.

Some coaches are great with Jira but when you ask them to switch to VersionOne, it all comes apart.

Some coaches excel in person-to-person coaching but struggle online.

Look for coaches that are tools agnostic and can help your organization apply good Agile and Scrum techniques with any tool you might use—in any environment your company operates in.

Think of coaches as writers. You have to be a good writer first; it does not matter if you write down notes using a pencil, type text into Word, or use some other editing program. It is the content that makes the writing valuable.

Coaches need to be tools agnostic.

Chapter 93
Coach Is Acting Like a Project Manager

Agile Coaches are not supposed to "manage" the teams they are coaching.

Unfortunately, some coaches go through the motions. They conduct some training, explain how things should be done, and then proceed to micromanage teams. Do this! Do that! Start this! Stop that!

That is not coaching; that is project managing.

Beware of project managers acting like coaches. You might end up with a good project manager, but you are not coaching your teams to become independent, self-managing, and empowered.

The hard part about coaching a team is that rather than playing on the field the coach needs to step back and let the team play.

The coach might not agree with how the team plays, but ultimately the team needs to own the game. The coach can give feedback at halftime. In the spirit of continuous learning, sometimes the best learning is based on failure—which is hard in corporate environments—but sometimes creates the best learning opportunities. Think about it. Have you learned more from your failures or your successes?

Chapter 94
Beware of Agile Coaches That Have Not Walked in Your Shoes

A doctor fills a vacant position as an emergency room surgeon in one of the busiest emergency rooms in a bustling metropolis. Would you trust that position to be filled by a doctor who has never been an emergency room surgeon? Say, he was a dermatologist for the last twenty years?

The answer is no.

Agile Coaches need to bring strong relevant experience to their coaching engagements. If you develop software, the coach needs to have a software development background. If you build skyscrapers, the coach needs a background in architecture or construction. If you run franchise restaurants, a prior background in running other franchise restaurants might be helpful.

Coaches that argue they do not need relevant experience usually argue so because they do not have it. It is a self-serving argument.

There are benefits to having coaches that have experience in various environments and industries because knowledge "cross-pollinates." But prior knowledge and experience in a specific domain are necessary to ensure good coaching results. Your coach should have walked in your shoes before.

Chapter 95
Scrum Masters Are Not Coaches

Coaches prepare folks to perform better, as individuals, as a team, and as a company. Coaches have a responsibility to coach with the interest of the business in mind.

Scrum Masters are part of the team.

Coaches are not part of the team.

Scrum Masters are more comparable to team captains in soccer. They play on the field. They are on the front line, and they provide in-play guidance and motivation.

Coaches never play on the field; they stand on the sidelines. They cannot intervene in the play except for the occasional encouraging word interjected between plays or when folks are swapped mid-game. Scrum Masters excel at removing obstacles and impediments for the team. Their focus is on the here and now. They are in the mix of it. Their focus is the team.

Coaches try to understand how to make individuals and teams perform better. Coaches excel at observing and advising. Their focus is long term. Their focus is holistic and includes the individual, team, portfolio, and company. Scrum Masters are not Coaches. Organizations that state as much are just trying to save money or provide career paths for existing managers.

Chapter 96
The Alarmist Coach: Everything Is a "Big Problem" or a "Mess"

How would you feel if you sat down on your therapist's couch and after a ten-minute conversation she said, "OMG! We have a big problem. You are a total mess!"

I do not know about you, but I would probably not visit again, and I would take my $240 and spend it on something else.

If you are working with an Agile Coach that communicates like this, in person, in email, or even worse, behind closed doors maneuvering for political angles, you have found yourself a book smart zealot!

They are usually easily identifiable by their signature line because they have a seemingly endless list of certification abbreviations intended to impress: John Doe, **MSAIT, SPC5, PMI-ACP, ICP-ATF, SSMC, CSPO, CEC.**

They are alarmists because the dirty reality of getting things done does not match the ideal agile process they have been taught. Crying wolf is also the ultimate CYA (Cover Your A..) move—in case things go wrong they will be the first ones to say, "I told you so!"

Beware of coaches like this; they do not provide a lot of value. Select coaches that lift you up, not tear you down.

Chapter 97
Agile Coaches Need to Know Their Obligations

Agile Coaches have two core obligations:

1. Inspire the professional potential in his customers and project teams. If you run the 100-meter dash in 22 seconds, the coach's job is to inspire you through various techniques to get you to run faster than that—according to your athletic potential. It is not his job to make you break the current world record of 9.58 seconds.
2. Coach with the customer's interest determining the direction. Every business is different. It is not up to an Agile Coach to determine how the business should be conducted—as such the coach needs to take guidance from the customer and coach accordingly. Too many times Agile Coaches come into organizations and quickly proclaim that this or that is wrong without fully understanding the customer's interests or direction.

Those two areas are at the core of what an Agile Coach does. Beware of coaches that do not look at this.

It is incumbent upon the Agile Coach to extract from the

customer exactly what the customer's interests and intent are—a good coach helps customers distill and articulate this information.

Part VIII
AGILE ANTI-PATTERNS

Following is a list of common **Agile Anti-Patterns.** This is mostly fun reading to illustrate what to avoid.

Chapter 98
When the Roadmap Does Not Match the Backlog

One of the more frustrating anti-patterns during an Agile or Digital Transformation is when the published roadmap does not even remotely resemble the Product Backlog.

Many times this simply indicates that the transformation effort might have started bottoms-up, with Product Management not fully aware how roadmaps should be presented in the Product Backlog—basically misalignment between strategic goals and tactical implementation due to a staggered transformation rollout.

Other times it might indicate a true disconnect between Product Management setting strategic direction and the agile implementation teams. For example, there might be a simmering disagreement between Product Management's vision of where the product should go versus a technical vision that does not align.

Remember, the User Stories in the Product Backlog are supposed to be detailed enough only for the next two or three sprints; beyond that point, the Product Backlog should rather focus on Themes, Epics, and Features from the product roadmap.

Chapter 99
Agile Process Does Not Drive Culture—Culture Drives Process

Peter Drucker famously said that "culture eats strategy for breakfast." The same applies to process.

Some people believe that when they establish a process, the culture will follow. It assumes a few learned specialists can guide the common workforce into agility and high performance by establishing a process that enables those desired characteristics. This is false.

The opposite frequently happens. Instead of enabling agility and high performance, the company just becomes more bureaucratic by rigidly following an Agile or Scrum process, without having enabled the critical culture part of the equation.

An agile process itself does not create cultural change. It creates a cargo cult.

If you want to change your culture, you need to address the company's vision, values, practices, people, and setting. If your culture is agile, your process will follow. Consequently, if you focus on culture first, your Agile or Digital Transformation will have a better chance of succeeding.

Chapter 100
Agile Confusion: Undergoing an Agile or Digital Transformation Is Not Just Another Tool or Tweak

It is a fundamental change.

Executives must realize that this one is different, for no other reason but the fact that your competitors that successfully transformed into agile organizations will outperform you by a wide margin.

It is easy to be misled into thinking that an Agile Transformation is essentially a technical concern. The reason it is not just technical is simple. Your value stream most likely consists of many other activities, departments, partners, suppliers, etc. that require synchronization and coordination—and your technical departments make up only a small piece of the overall value stream.

If you only make the technical side of your business more agile, you will simply not realize the advantages that you are hoping for. You are optimizing a small percentage of your business.

It is incumbent upon senior executives to realize that if the benefits of an Agile or Digital Transformation are indeed to accrue, then change must be demanded from—and sponsored across—the whole enterprise.

Chapter 101
Nine Steps to Agile and Digital Transformation Hell

Here is one of the most common anti-patterns for large-scale, corporate, Agile, or Digital Transformations:

1. Executives hear about Agile and decide that this will save the company.
2. The Company hires a big-name consulting company to implement the Transformation Program.
3. The Consulting company creates a 390-page PowerPoint presentation "Playbook".
4. Management is sort of, kind of, trained on Agile but old reporting structures stay in place.
5. Staff is sent to a 2-day class and obtains a "certificate" testifying to their expertise.
6. Some new talent is hired with previous Agile experience (and similar 2-day class certifications).
7. IT rolls out Jira, Jira Portfolio, VersionOne, or another tool but the staff is not sufficiently trained on the tool.
8. The big-name consulting company leaves, claiming another "Transformation Success"—executives claim victory.
9. The existing PMO is renamed into Agile PMO and starts

micromanaging the Transformation Program.

Result: Things look suspiciously like they did before the Agile or Digital Transformation started. Business value is still not delivered any faster. Customer satisfaction has not improved. Financial improvements have not materialized. Shareholder value has remained the same.

What a surprise!

Chapter 102
Not Hiring the Right People

Agile Transformations are like permanent weight loss—they signify a fundamental change in how the company operates and how employees think, work, and act.

Just like diet fads that promise quick results, many companies chose to run an Agile or Digital Transformation effort like a PMO program—achieving short term results (i.e. quick weight loss) but reverting back to old habits before the company is truly transformed (i.e. maintaining permanent weight loss).

In large companies, you can usually observe this anti-pattern over a three to five-year time span during which a big transformation gets rolled out and somebody declares victory: We are now agile!

Ultimately, the proof is in the pudding.

- Are you delivering value to your customers faster?
- Are you achieving a constant flow of value?
- Are employees actively collaborating on problem solutions or just passing around documents?
- Are you delivering a better quality product?

Beware of Agile and Digital Transformation efforts that are run like a PMO program.

Chapter 103
Not Hiring the Right People

There is only one thing you need to do to succeed. I know what you think. One? Really? Just one? We are doing twenty-nine things and are still failing! Learn to hire the right people for the positions you are filling. Know what you are hiring for. Always. If you do not know what you are hiring for, stop! Figure it out before proceeding!

Be aware of where your company is in terms of company lifecycle. Startup? Innovator? Early adaptor? Mass market? B2B? Does the potential hire fit? There must be a fit or you create problems down the road.

Hire the right people, and your Agile problems will go away.

Hire the wrong people, and your staff will become the proverbial ball and chains dragging down the company's performance and success.

At its core, Agile is about people—what they know, what they care about, what they want to achieve, and how they drive things.

Get good at hiring the right people, and all the other problems solve themselves. On the other hand, get the wrong people and all the Agile and Digital Transformation initiatives in the world will not help you. This is the linchpin to Agile success.

Chapter 104
Unable to Hold Two Thoughts at the Same Time

Some folks believe **Agile** and **Scrum** will solve every problem. Because everybody is striving for certainty, many people stop questioning the complexities of product development, assuming that Agile and Scrum will be the solution—forever.

But certainty does not exist. New technologies, new developments, outsourcing, offshoring, regulatory changes, competitive pressures, and pandemics happen—nobody knows where things are going. Anybody pretending to is just fooling themselves.

Some megatrends will directionally point the way: Internationalization, international labor arbitrage, remote work, global warming and rising sea levels, automobile electrification, China as the largest economy, etc.

But you cannot buy into the Agile hype believing that it will solve all problems. Agile and Scrum are temporary and will require serious rethinking as we go along. What works today will not work in thirty years.

Critical thinking will still be required; try to hold two or more thoughts in your mind or anticipate failure! Only the paranoid survive!

Chapter 105
The Foolishness of Agile Certifications

The **Rashomon Effect** describes how people describe an event or situation in different and contradictory manners, which reflects their subjective interpretation and self-interested advocacy, rather than objective truth.[1]

Samples of this include notoriously unreliable eyewitnesses or parties of divorce proceedings ignoring actual events or the interests of their children.

Based on people's vantage point, background, experiences, interests, and education proponents of Agile and Scrum will argue vehemently that one approach or the other, one technique or the other, one methodology or the other, makes the most sense.

This is usually influenced by their self-interest and preferred focus areas, not by what is objectively correct or makes the most sense for the client.

Even worse, sometimes there are flat out disagreements by Agile experts as to what the problem or the potential solution could be.

Beware of the Agile Rashomon Effect—rational and objective decisions are hard to come by.

Chapter 106
Agilists and the Elephant in the Dark

When talking to fellow **Agilists** or **Agile Coaches**, I usually try to understand their viewpoints first before commenting on their statements. My background is mostly high tech, from the early days of hardware development over UNIX/LINUX development to iOS/Android and lots of Enterprise Software in between. Everybody brings their own perspective and experience.

The parable of the blind men and an elephant comes to mind. It originated in ancient India. It is a story of a group of blind men who have never come across an elephant before and who learn and conceptualize what the elephant is like by touching it.

Each man feels a different part of the elephant's body, only one part, such as the ear, the leg, or the tail. Then they describe the elephant based on their limited experience. Their descriptions differ widely.

Some come to suspect that the other person is dishonest, and they start arguing.

The moral of the story is that people tend to claim absolute truth based on their limited, subjective experience.

Welcome to the Agile Methodology Wars!

Chapter 107
The Foolishness of Agile Certifications

IF

LinkedIn profile or business card has a long list of certs LIKE (CSM, CSPO, ICP-ACC, ICP-BAF, SAFe Agilist, MSAIT, SPC5, ICP-ATF, SSMC)

THEN

Insecurity is off the charts!

Let's not pretend that two-day certifications are the same as an MD.

They are not! And everybody knows it. Help yourself and your credibility by not putting them on your profile or business card!

Chapter 108
The Top-Heavy Team

This anti-pattern addresses "top-heaviness," meaning the team has too many senior-level leaders but not enough lower-level working engineers.

Sometimes one can observe this due to junior team members "aging up," meaning they joined the team as a junior member but after several years of experience have become more mid-level or senior level.

Sometimes this pattern can be observed because management is concerned about the product/project, meaning all senior-level expertise available is allocated to one project because it is perceived to guarantee success.

Please note that seniority by itself is not the problem; the challenges come in if senior level engineers do not want to perform what is perceived as simpler, grunt level, engineering tasks usually assigned to junior engineers—that is when dysfunction sets in and tasks do not get done.

As long as senior-level folks are willing to work on whatever is necessary to get the job done, this is fine. But in the long run, some dissatisfaction usually sets in. This "top-heavy" model ultimately is not sustainable.

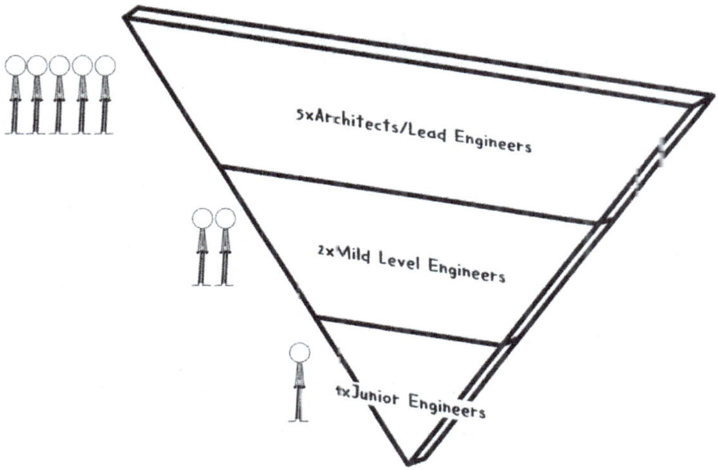

Figure 11 - Top Heavy Team Pattern

Chapter 109
The Inexperienced Team

This common team anti-pattern is the reverse of the Top Heavy Team pattern; it essentially deals with too many junior-level folks on the team who do not get enough guidance, training, and mentoring from more senior team members.

Simply put, the Agile/Scrum team overall is too inexperienced.

This pattern is common when managers staff an Agile/Scrum team "to budget," meaning that budget limitations cause the hiring manager to put in too many junior-level folks without regard to what is needed from a team composition perspective. It can also be caused by a resource shortage based on competing programs within the company consuming all senior-level staff or a myriad of other reasons.

The bottom line is if you come across this kind of inexperienced team, you either need to rebalance the team with some more senior resources or you should seriously reconsider if you want to kick off the effort like this. If in doubt, it is better to delay than to get started on the wrong foot.

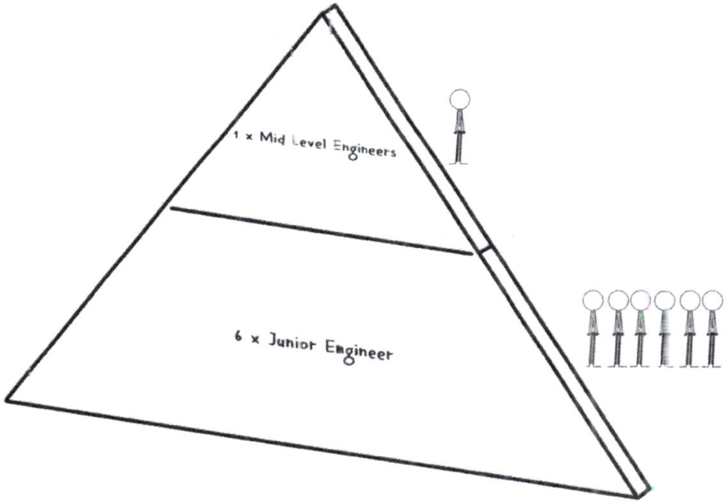

Figure 12 – Inexperienced Team Pattern

Chapter 110
The Geographically Dispersed Team

A common team anti-pattern deals with the extreme **geographic distribution** of an Agile/Scrum team.

There are different kinds of geographic distribution; some are worse than others.

The best or the easiest kind to deal with is geographic distribution within one time zone. The current COVID-19 crisis has forced many employers into this situation.

The worst kind of geographic distribution for an Agile development team is extreme distribution across many time zones and different cultures.

Extreme geographic distribution can happen for many reasons, such as:

- Skill sets are not available locally.
- Skill sets/domain expertise is only available in other company locations.
- The company is geographically distributed and wants to use available staff members in other locations.
- Budgets force the hiring manager to build a virtual team with team members in countries with lower labor costs.

- Part or all of the development effort is outsourced to partners in different locations.

No matter the reasons, the challenges of team distribution can be overcome with a focus on technology (video, voice, text), a disciplined process, and maximum flexibility. This is difficult but doable. However, this complicates an Agile Transformation tremendously.

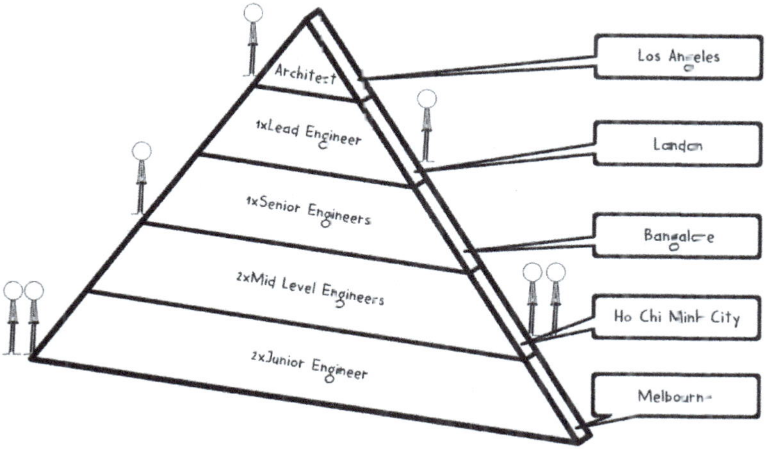

Figure 13 - Extreme Geographic Distribution Pattern

Chapter 111
The Stagnant Team

Agile/Scrum coaches constantly mention "stable teams." Of course, the intent is to establish a team and provide the environment to truly build expertise, working relationships, cadence, and process improvements.

Teams that know each other well, deliver better products, in shorter time frames, and higher quality. In many ways, Agile/Scrum assumes stable teams as the basis for productivity improvements and faster value delivery. No surprise here.

But, just like in families, long-established teams develop certain personality dynamics. There is dad; there is mom; there are the kids, and occasionally crazy Uncle Willie comes by.

Long-established teams run the risk of stagnating.

The point is that although stable, long-established teams are not necessarily bad, you need to look out for people getting complacent in their roles. Sometimes bringing in new blood helps invigorate a team and provides new ideas.

Stable is good. Stagnant is bad. There is a reason that the military rotates troops off the front lines or sports teams replace players. Agile/Scrum teams need to do the same.

Chapter 112
Building Up Technical Debt

Traditional development efforts focused on getting through a set of phase gates (analysis design > implementation > testing >deployment). The purpose of phase gates was to show progress.

This resulted in developers often having a party when the implementation phase was done—"Coding is done, now it is up to the test team!" In reality, development ended up with a mountain of bugs; even worse, because testing had been identified as a phase, nobody even knew how many bugs were in the system. Endless break/fix cycles commenced. Technical Debt had spun out of control.

Agile and Scrum processes introduced empirical process control, backlog management, the Definition of Done concept, and close stakeholder involvement via the Product Owner.

Modern development techniques introduced trunk-based development, automated unit testing, behavior-driven development techniques, KPIs, and the standardization of containerized or virtualized environments. All this shifted the testing phase to the left and made the true state of the system visible at any given time.

All this will not prevent you from building up technical debt, but now we have the tools to know exactly how to keep technical debt to a minimum.

Chapter 113
Slicing Your Backlog Based On Component

If your **Product Backlog** items are sliced horizontally based on components (UI, Business Components, Database, or Infrastructure) instead of vertically based on end-to-end features, you have a problem.

The challenge here is that the root cause could be one of many:

- Your team is not truly cross-functional.
- Your organization might have old, established silos.
- Your team is used to thinking in component terms (because of silos).
- Your Product Owner is not talking about end-to-end features.
- Your Product Owner might be too technical and not focused enough on the customer.

Thinking vertically, end-to-end reinforces a customer-centric view. Your focus is on **what** has to be produced.

Worrying about components indicates that you are concentrating on the internal, non-value-adding activities. Your focus is on **how** things are produced.

Product Owners (who are responsible for the Backlog) need to focus on the "**What**?" Let the Scrum team focus on the "**How**?"

Chapter 114
Not Involving the Team

If you are not involving your team in planning, defining, and sizing of work, you are not Agile.

Only having leads and managers independently plan things for the teams negates the entire idea of Agile.

You might as well stick with your Gantt charts.

Chapter 115
Not Having a Definition of Ready (or Definition of Done)

This anti-pattern comes in two flavors:

1. Not having a **Definition of Ready (DoR)** or **Definition of Done (DoD)** or
2. Not sticking to your own **DoR** or **DoD**.

The latter is worse than the former because you went through the effort of creating a DoD or DoR but you are not following your own rules!

For a more in-depth review of the Definition of Ready versus the Definition of Done, see my article on LinkedIn.[1]

Chapter 116
THE PRODUCT OWNER PROXY OR PRODUCT OWNER AGENT

The most important role in Agile/Scrum is the **Product Owner**.

With a strong **Product Owner**, the team rises. With a weak Product Owner, even the best team has a big chance of not delivering.

Think of the Product Owner as the quarterback in football, touching the ball on every offensive snap, and being the triggerman for the passing game. Or, in soccer, the defensive midfielder, knitting together the plays between attack and defense.

Product Owners weave business needs and delivery capabilities together.

The Product Owner is the only person to decide which tasks become Backlog items and decides on priority by ordering the Backlog.

A common challenge in large organizations or companies that use extensive outsourcing is that the Product Owner Proxy or Product Owner Agent has to coordinate and get approval from stakeholders because she is not empowered.

This coordination takes time, is prone to misinterpretation, leads to lengthy back and forth, and ultimately leads right back to a waterfall-ish development approach.

If you can only influence one thing, get a great Product Owner! For more about what makes a great Product Owner, read my article on LinkedIn.[1]

Chapter 117
No Slack Time

Agile is often-times misunderstood as a way to squeeze productivity out of your teams. In some people's minds, productivity means you work your teams at 100% utilization. It is a very industrial revolution mindset. Some folks even apply a word, fungibility, to describe how you can plug in workers wherever and whenever you need them, guaranteeing 100% utilization with no slack.

In Agile/Scrum, teams need to have **slack time**—a recommended allocation is 20% to 25%.

Running at 100% utilization will make your products worse, not better. Team performance drops because individuals only focus on getting their specific items done with no regard to how their teammates are doing. Pairing or partnering on problems becomes impossible. Constant rework will become necessary as items are rushed to completion prematurely. Everybody is busy, but less gets done.

Again, without slack time, productivity drops.

For a brilliant book about this subject, check out Tom DeMarco's *Slack: Getting Past Burnout, Busywork, and the Myth of Total Efficiency.*[1]

Chapter 118
When the Sprint Review Turns into a Political Showdown

Beware of stakeholders and other influencers in your organization who may use the sprint review meeting as a battleground for a political showdown.

Good **Scrum Masters** and **Product Owners** know how to effectively navigate a situation like this, and to make sure everybody knows the rules. Ideally, both the Scrum Master and the Product Owner have talked to all stakeholders and influencers beforehand to set expectations.

It is important to make sure that stakeholders attending the sprint review meeting understand that they can listen and observe, but they cannot abuse the meeting to carry out political hit jobs on the project.

The development team should not get exposed to whatever political/strategic/budgetary disagreements may be brewing behind the scenes, as it causes the technical staff to get disillusioned and question the leadership.

The focus of the sprint review meeting is just that—to review the sprint and the associated deliverables, the product increment. Keep other discussions out of it.

Chapter 119
Waterfall Planning

This is one of my favorites. After carefully getting your Agile/Scrum teams trained up and prepared for that big Agile at Scale implementation effort, the teams proceed to create Product Backlog items encompassing the entire five-year project schedule and assigning user stories to all 180 planned sprints. To top things off, the teams "baseline" their sprint scopes for all 180 sprints.

That is not the way it works. Beware of doing Fake Agile. Just labeling something Agile but executing it in a Waterfall pattern does not make it so.

If it looks like a duck, swims like a duck, and quacks like a duck, then it probably is a duck.

Chapter 120
Focusing on Process over Intent

The intent of agility is to provide business value—cheaper, better, and faster—at a sustainable pace.

If you spend all your time documenting step-by-step prescriptive processes in endless policy documents, you are wasting your time.

Focus on intent, not process!

Chapter 121
When the Product Owner is Surprised About the State of the Increment

Product Owners are supposed to be dedicated to the development team and should interact with the team daily.

However, the reality is that many Product Owners have "day jobs," and as such, often do not have the bandwidth to participate in all Agile/Scrum events and meetings.

Sometimes, Product Owners are "out of the loop" on the latest coding challenges and this can lead to the Product Owner being surprised as to what functionality was finished or not when participating in the sprint review meeting.

To avoid this, make sure that:

1. Ideally, the Product Owner is dedicated full-time.
2. Ideally, the Product Owner attends all necessary Agile/Scrum coordination meetings (sprint planning, daily scrum/standup, sprint review, and sprint retrospective).
3. Functionality is developed throughout the sprint, avoiding a big wave of code to be finalized shortly before the sprint review.

The scrum board is updated with the latest information, so the

Product Owner can review any time; it is a good practice to start the sprint review meeting with a quick review of the board.

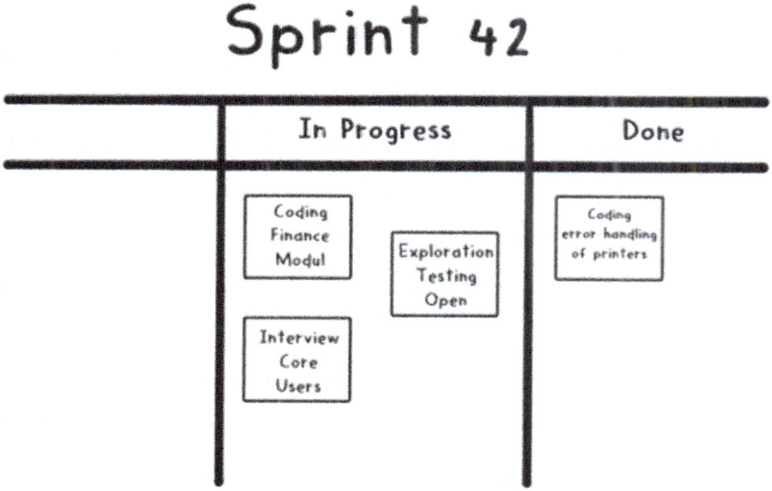

Figure 14 – Updated Scrum Board

Chapter 122
No Research Activities

If your Product Backlog contains no or only very few research spikes, your team might be stuck in perpetual problem-solving mode—meaning your team has endless discussions about problems, instead of researching them with a spike as a part of an iterative item creation process.

Beware of your team falling into "solutioning" discussions all the time.

There are problems you know how to solve, size, and estimate. Then there are problems you need to do research for—simply because you do not know how to best implement the requested functionality.

Having no research spikes usually indicates that you are stuck in "solutioning."

Chapter 123
Missing Acceptance Criteria

By design, the Product Backlog is a moving target. It s not a static picture. The Product Backlog morphs over time—items can be added, updated, and deleted until you want to assign them to a sprint.

At the beginning of the story refinement cycle, your Product Backlog items might not have acceptance criteria. There is no specific point of time prescribed in the refinement cycle when acceptance criteria have to be done.

But, the Definition of Ready (DoR) should clearly state the need for acceptance criteria—because acceptance criteria provide implementation boundaries. They make the implementation tasks more real, more manageable, and provide a clear definition of success. So, acceptance criteria will have to be defined somewhere between the time of the creation of the user story and the time it is pulled into a sprint. A team member cannot work on a user story without it having acceptance criteria defined and agreed to by the Product Owner.

If user stories in your active sprint do not have acceptance criteria, you are in trouble because you most likely will not be able to judge success or failure—and might miss your sprint goal.

Chapter 124
Product Backlog: Too Much, Too Little, Not Enough

Too Much: The team is constantly refining, into ever greater detail, to the point of having a too detailed backlog. I have seen user stories with pseudo code, detailing the coding logic the programmer had to follow—I kid you not!

Too Little: The team does not have enough refinement sessions; they never catch up to the point of having two to three sprints worth of good user stories. Because the team spends too little time doing thorough refinement sessions, the user stories do not provide the guidance to the team that is needed. This in turn does not allow for effective sprint planning, so the team blows their sprint goals regularly.

Not Enough: The team regularly meets for refinement sessions, but never gets sufficiently detailed enough. User stories are too general, using big generic words that can be interpreted in many ways, with no or insufficient acceptance criteria, undiscovered dependencies, etc.

All these variations can result in a low-quality backlog. Spend concentrated, quality time on your refinement process or pay the price in low quality deliverables and disappointed stakeholders downstream.

Chapter 125
Unrealistic Ideas

Agile and Scrum are about planning and delivering for the here and now. We acknowledge that we cannot predict the future and we understand the concept of the Cone of Uncertainty.

I have seen Backlog entries for startup teams (with startup money or lack thereof) that talked about world-wide domination of their market—without having one line of code written.

Big ideas are great, but Backlog items need to be believable, achievable, and realistic. Keep big ideas that might materialize in five years separate from actionable Product Backlog items that your teams can tackle now.

It is OK to dream big, for investors to see the potential, or for your boss to see where you want to take things. But the portfolio plan, release plan, and specific Product Backlog must be believable, achievable, and realistic—otherwise, you risk your teams wasting their time.

Chapter 126
Copy and Paste from Business Requirements Document

A major challenge for existing companies that are going through a Digital Transformation is that life before Agile produced a stream of artifacts, in formats that have nothing to do with the standard User Story format.

Most of the time there is deep historical knowledge documented in Business Requirement Documents, Functional Specifications, or similar types of documents, so they simply cannot be discarded.

Now what? For legacy systems that are already in production, let things be. Do not try to retrofit existing requirements into a Product Backlog and a User Story format.

For new development efforts, those that are partial replacements of existing modules, or replacement efforts for entire systems, follow the standard User Story format in your Product Backlog. But, do not just copy and paste from existing documents.

The purpose of writing User Stories is to go through the "card-conversation-confirmation" process; as such, the process of writing User Stories is a team exercise, for creating common understanding amongst all team members. Copy and paste will not do that.

Chapter 127
FORGETTING ABOUT OR NOT UNDERSTANDING INVEST

User stories are supposed to be:

Independent (of other stories).
Negotiable (not a specific contract for features).
Valuable (Why do something that is not?).
Estimable (to some approximation).
Small (so it fits in a sprint/iteration).
Testable (do not write stories that cannot be tested).

It is always worthwhile remembering time-honored classics, like this early post by Bill Wake, "Invest in Good Stories, and SMART Tasks."[1]

Chapter 128
Not Understanding Walking Skeletons

The idea of a walking skeleton is simple: Provide basic, core functionality first, demonstrating the main elements of the design before fleshing out bells and whistles. "Steel Thread" is another term that describes this.

Reflecting this in your Product Backlog and layering even more detailed user stories across sprints or Program Increments can be tricky, specifically for folks that have come from traditional waterfall efforts. Why? Because people have a natural tendency of trying to think through the entire product space instead of acknowledging "unknown unknowns" and relying on progressive experimentation.

We know that planning out everything ahead of time does not work—we simply have no idea what challenges, problems, dependencies, or staffing changes we will experience in the future.

So, the trick to creating a good Product Backlog is to work iteratively and incrementally.

You iterate towards your calendar target, while incrementally improving your product.

For example, look at the 1980s Toyota Camry as a walking skeleton—and look at the current model Toyota Camry now.

This is yearly model iterations and incremental product improvement at its best.

Chapter 129
When Your Service Provider's SOW Contradicts All Things Agile

In large, complex, system implementation efforts you are frequently faced with just as complex organizational structures. You might have a combination of full time employees, contractors, and consulting firms, in multiple locations, all working on a large scale program.

In an ideal world, you would organize your Agile/Scrum teams, manage the work through Kanban or Scrum boards, and happily deliver business value in frequent releases. Done.

In reality, one of the biggest obstacles to agility is when your company retained one or multiple service providers that are working on Statements of Work which have been structured along with traditional waterfall deliverables—hard date commitments, guaranteed staffing levels, penalty clauses, etc.

In other words, you have this big waterfall square you are trying to fit into your round Agile hole.

My advice? Next time around, do not do this again; negotiate the SOW based on business value delivered and goals achieved, not based on paper-based plans that might change. Embed your consulting partner into your Agile process.

This time around, you will have to work around issues and try your best.

Chapter 130
The Confusion about Scrum, Agile Methodologies, Techniques, and Practices

Here is a cheat sheet:

1. **Scrum**—at the core you will hear about Scrum, which is simply a team collaboration framework. If you want to talk about the basics, you need to talk about Scrum.
2. **Agile Scaling Methodologies/Frameworks**—these are usually addressing the end-to-end application of Agile and Scrum at scale in enterprise environments. How do you use Agile and Scrum across dozens of teams? The usual suspects include SAFe, DAD, Nexus, and LeSS.
3. **Techniques**—there are specific techniques that are associated with Agile and Scrum, such as making work visible using Burndown or Kanban boards.
4. **Practices**—there are common best practices, such as unit testing or build automation that are used during software development. Many practices are not specific to Agile and Scrum and could be applied to any SDLC.

DevOps is sometimes mentioned as a methodology, which is wrong. DevOps is a set of techniques to deliver software to production. Beware that if you ask ten agile experts, you will likely get fifteen different answers.

Chapter 131
Too Many Cooks in the Kitchen

Agile and **Scrum** describe a simple responsibility model: One team, one Product Owner, one Scrum Master. SAFe extends that to one RTE per Agile Release Train.

One person responsible. This is generally acknowledged to be the most effective responsibility model. Yes, there are sometimes exceptions, but normally one person responsible is the way to go.

Introduce two Scrum Masters, two Product Owners, two architects, or multiples of any other role, and suddenly you have committee meetings. Productivity drops, conflicting opinions appear that need to be reconciled, and clarity vanishes.

All of this assumes competency on part of the person filling a particular role. If you do not have a competent candidate to fill the role, find one.

Remember two things:

One, understand the "real" organization—you might only have one person with the title, but where does the real power lie?

Two, simplicity rules—simple organization, simple team setup, simple architecture. Too many cooks in the kitchen make for complex decision making, complex solutions, and complex maintenance.

Chapter 132
When Measures Become Targets

Measures are good.

They allow teams to understand what is going on, and to self-calibrate. For example, if you think about how to improve a teams' soccer game, the team might be interested in their pass accuracy or number of shots on goal. This is useful information for the team to improve their performance, switch players, change their training routine. It helps the team.

But when measures turn into targets, things go bad.

Probably the most common anti-pattern to showcase this is when folks look at velocity and do not understand that it is a team measure. Too many companies turn velocity into a target, and suddenly teams are judged because they did not hit a velocity of "x." This hurts teams and is counterproductive.

Targets invariably cause problems and do not improve performance. Measures, used by teams, do.

Chapter 133
Not Providing an Ideal Office or Remote Setup

An Agile or Digital Transformation requires reconfiguring your workspace to match the Agile/Scrum process. This sounds simple but can be challenging. Facilities need to consider the optimal office layout, conference rooms, break rooms, rest areas—in general, the best appropriate physical environment.

In the virtual remote world this means the top of the line connectivity, high-speed connections, and cross-platform productivity software—essentially you need to be able to be connected all the time, on any platform—laptop, desktop, phone, tablet.

If in an office or remote, optimizing communication while maintaining a quiet work environment is the devil in this detail.

With the current pandemic, setup considerations necessitate thinking through onsite, remote, and hybrid scenarios.

The sad truth is that many companies miss the boat on this one because they assume one or the other future—all office or all online—but the reality will probably be some combination of real office presence in combination with a strong remote work component.

Welcome to the new normal!

Chapter 134
Not Using Personas in User Stories

New Agile teams that are trying to write user stories for the first time frequently miss the boat when it comes to personas. You can easily see this because every user story starts with "As a user, I want …" What user?

You do not care about a generic user, what you really need to understand for good user stories are the relevant User Personas.

A **User Persona** is a fictional representation of a group of users of your application/app/website.

Personas are based on available information from Product Management, user research, stakeholder interviews, site analytics, and brainstorming sessions. Defining User Personas is usually a team effort.

How? Standard User Persona elements usually consist of:

- Name (Fred, Laura).
- Photo (stock photo or draw a picture).
- Age, Gender, Marital Status.
- Education Level.
- Job/Work Experience.
- Location.

- Comfort with technology.
- One sentence description of this user.

Defining several User Personas that represent the real users of your system will enable you to write better user stories that have real meaning and provide better information for the implementation team.

Part IX
AGILE BEST PRACTICES

Best Practices are frequently a discussion topic that brings out passionate arguments. After all, what is a best practice anyway and who decides?

Wikipedia provides a good definition: "A best practice is a method or technique that has been generally accepted as superior to any alternatives because it produces results that are superior to those achieved by other means or because it has become a standard way of doing things, e.g., a standard way of complying with legal or ethical requirements."[1]

Alternatively, Merriam-Webster defines a best practice as "a procedure that has been shown by research and experience to produce optimal results and that is established or proposed as a standard suitable for widespread adoption."[2]

The best practices listed in this chapter are entirely based on my work experience in the agile consulting world, and specifically in large scale software development and IT systems development.

Chapter 135
Leaders Are Educated, Not Trained

When I ask how to get executives on board with an **Agile or Digital Transformation**, people say things like, "We have to train our executives to look at things differently." In my experience, you do not "train" them, you must educate them.

You do not sit down with a PowerPoint presentation and lecture your leaders how to look at the world—they have their own views, experiences, expectations, and styles—and, most importantly—they are where they are for a reason.

Training equalizes participants because you teach tools, techniques, and skills. Afterward, everybody has the same knowledge.

Education, on the other hand, makes people different because it takes into consideration their position and the personal experience of the individual while introducing them to new ideas.

Educate them! Hallway conversations, quick informal discussions, informal lunch meetings, helpful emails with interesting and concise information that is relevant go a long way in getting your leaders to change their view.

Talk to your leaders and executives, do not talk at them.

Chapter 136
Product Roadmaps Are Funded at the Epic or Feature Level

Product Roadmaps are represented in Epics and Features or any other issue type hierarchy.

The important thing is to remember that the issue type hierarchy is simply a tree structure that allows for the functional decomposition of large bodies of work into smaller digestible chunks.

What each level is called depends on the agile tool you use, and what agile framework you follow. The lowest roadmap level is finally decomposed into User Stories, which are actionable pieces of work that Agile teams can implement and deliver.

The highest level (Epic or Feature level, depending on what tool you use) usually represents budget items or budget categories that represent real money allocations.

Remember that you need to tie together your Agile business goals with your budgets—otherwise, you are just pretending.

Chapter 137
Do Not Start an Agile Transformation Without a Clear Goal in Mind

Any coach in any discipline (be it soccer, golf, tennis, basketball, success management, financial management, college prep, or business level coaching) knows that the first step is to understand the *desired results of the transformation*.

What do you want out of the Agile or Digital Transformation? "Being Agile" is not a goal, just as "being thin" is not the goal of a diet.

When starting an Agile Transformation, pick no more than three goals!

Why? Because picking 5 or 8 goals is the equivalent of quitting smoking, losing weight, starting a new job, getting a new baby, moving, and taking care of your elderly parents all at the same time. It is not going to happen!

Chapter 138
Continuously Monitor Everything

Modern development techniques allow for continuous system monitoring at all levels—all the time!

During the development cycle—Developers monitor their unit test coverage. Security tools scan for potential risks in their code. Tools can automate the build and deployment process and monitor if, when, and how builds or deployments fail. Automated tests can be triggered every time a build is run and monitor how many tests pass or fail.

In production—System/server uptime is monitored as a matter of guaranteeing customer satisfaction. Development teams now instrument their code to monitor how the features are being used in production. We monitor how many users are logged on, how many transactions happen, how many are successful, or how many are abandoned. We monitor who clicks on what, when, how.

All this info being gathered is used to assess, prioritize, and optimize the development process, the production environment, and the users themselves.

Ultimately this information is used to create or help prioritize backlog items, that feedback into the development team.

If you do not monitor things, you are flying blind.

CHAPTER 139
VELOCITY TRACKING

Tracking the velocity of one or many teams should be standard practice for scrum masters, release train engineers, and team members. It is pretty straightforward.

Calculated, based on story points:

- Rolling average of sprints.
- Maximum.
- Minimum.
- Lifetime average.

Used to:

- Determine sprint scope.
- Calculate the approximate cost of a release.
- Track release progress.

Remember these best practices:

- Velocity without a quality metric is not useful.
- Velocity of two teams is not comparable.
- Velocity changes when the team composition changes.
- Velocity increases with a team's tenure.

- Velocity does not equal productivity.
- Defects and rejected user stories count against velocity!

Remember, any metric gathered can be abused and used against the team—so make sure that does not happen, otherwise, the velocity metric becomes useless.

Chapter 140
Team Turnover Tracking

Frequently people look at their team velocity or the velocity of several teams in a program increment, and they are puzzled when velocity numbers are all over the place. Sometimes teams hit 30 points, sometimes 60, sometimes 20, and then go back to 33.

There are lots of legitimate reasons for velocity to fluctuate. Vacations, holidays, sick time, company events, etc.

However, in my experience, the number one culprit when you see velocity numbers going all over the place like this is simple: staff turnover.

From a company perspective, turnover is the act of replacing an employee with a new employee, whatever the reason. If your company uses staff augmentation, you will also need to consider contractors and/or consultants in this number.

Whatever the reason, turnover is the bane of consistent velocity.

Many industries deal with an average of 10% to 15% turnover. So, if your Agile/Scrum teams have a turnover of 30% or 40% (which I have seen), you will never, ever, realize consistent velocity numbers. Or, even worse, you might get used to seeing very low numbers. Stable and improving velocity numbers are based on stable teams, i.e. low team turnover.

Chapter 141
Use Risk-Based Test Prioritization

In a deadline-driven delivery environment, the essence of testing is to execute as many relevant tests as possible in as little time as feasible.

"As many relevant tests as possible" is bound by how fast your test automation can run and log results. "As little time as feasible" is considering your environment, budget, and business needs. Instantaneously running automated unit tests every time somebody checks in code is now commonplace. Running frequent automated integration, functional regression, and load tests is also standard practice.

Despite the best Definition of Done, if a problem shows up after the fact, then the first question that is asked is "Why did our tests not catch that?"

Consequently, you need to understand and vigorously use risk management and prioritize your work using Pareto analysis (aka the 80/20 principle). 80% of defects originate in 20% of code.[1]

Understand the root causes of your defects, and you will have a better chance of prioritizing your relevant test automation work.

Development techniques, static and dynamic code analysis tools, and system monitoring all provide you with the data to prioritize your tests using Pareto analysis.

Chapter 142
How to Progressively Stabilize the System—Automated Unit Tests

All test automation depends on progressively stabilizing the system, and the basis for this is automated unit tests.

Automated unit tests are the foundation upon which other automated tests (integration, functional, system, regression) are written. As a matter of fact, I have seen some development efforts drive their validation activities entirely through automated unit tests—it depends on what you are developing.

Too many times functional test automation efforts, targeted at the UI level, fail because they try to automate at the system test level without having the corresponding unit tests automated in the first place.

System instabilities and inconsistencies rattle through the system, and hit automated UI tests especially hard, causing lots of rework and frustration.

That is usually a sign of a big bang test automation effort at the end of the development cycle and spells doom.

If your teams are not implementing automated unit tests, do not try to automate functional testing–or put the appropriate focus on unit testing.

Chapter 143
Experimentation – Plan – Do – Check – Act (PDCA)

Learning by experimentation is the best way for you to figure out what works and what does not. After all, that is how we all learned to crawl, walk, and run.

That is how scientists frequently figure out how things work. Plan something, do it, check the results, act on the finding. W. Edward Deming did it; Toyota does it; so can you.

Do not be afraid to change things up to see if things work better one way or the other. What do you have to lose?

This applies equally at all levels of the organization–from the portfolio and program level all the way down to the team level and individual contributors.

Figure 15 - Plan - Do - Check - Act

Chapter 144
Remember Software Development Basics

As Agile/Lean practitioners, we frequently forget about the time-honored basics that have made software development probably more predictable than all process innovations combined.

Too often we focus on complex processes instead of concentrating on software development basics. Here are my favored two.

1. Keep It Simple Stupid (KISS) - most systems work best if they are kept simple rather than made complex. Complex solutions are difficult to build, hard to maintain, likely less stable to operate, and impossible to debug if problems occur. Simple is good.
2. You Aren't Gonna Need It (YAGNI) - do not add functionality until absolutely necessary. In other words, do not start adding functionality that you think you might need in the future, because we are all bad at predicting what is to come. Focus on the needs here and now.

Lastly, remember that the "Cone of Uncertainty" and Lean principle of "Decide as late as possible" are related to both KISS and YAGNI.

Chapter 145
Understand the Limits of Root Cause Analysis

Root cause analysis and fishbone/Ishikawa diagrams are pretty much standard practice nowadays–but they are losing their effectiveness.

This is because problems are increasingly complex, and when analyzing problems, you are faced with multi-dimensional and frequently multi-disciplinary problems.

For example, think about the complex machine learning algorithms that can be impacted by faulty logic, bad data, location-specific dependencies, insufficient sample data when creating the model, or a set of many other factors.

The days of asking "why?" five times to get to the bottom of a problem are long gone.

When faced with these kinds of problems, think about putting together a multi-disciplinary team and swarm over the problem in short, concentrated, work sessions. I have found this to be shockingly effective in quickly analyzing and identifying highly complex problems. The key is to have participation across all disciplines.

Figure 16 - Root Cause Analysis

Chapter 146
Agile Test Automation: Keep Developers, DevOps, and Test Engineering in Sync

Standardize on one programming language and one development environment (IDE).

It is a bad idea to have the test engineers work in a programming language (say Python) that is different from what the other software or DevOps engineers use for the project (say Java).

If your project or product is mainly written in Java have the DevOps and test engineers use Java as well (meaning have them use the Java language bindings for Cucumber, Selenium, etc.). If your product is developed in C#, then have them use C#. The same applies to the IDE; have all engineers use the same. Why?

The logic for this is threefold:

1. When people assist each other, it is good to have one language as the common base.
2. It allows for mobility within the team and encourages people to help each other; folks will not get stuck in one job because they only know a specific scripting language.
3. It makes integration of various tests easier (for example, unit test, web service tests, and functional tests all being written

in Java are easier to integrate into the continuous integration setup).

Simple choices like this improve team productivity.

Chapter 147
Develop Iteratively

If there is one best practice that has established itself without much discussion it is this one: **Develop Iteratively**!

Most modern software development projects use some kind of spiral based methodology rather than a Waterfall process.

There is a myriad of choices out there, from older legacy models such as the Rational Unified Process (RUP), to the last Agile/Scrum, SAFe, and Extreme Programming (XP) frameworks. Developing iteratively follows the long-established Deming Cycle (Plan-Do-Check-Act). DevOps techniques have enabled almost instantaneous feedback loops to the software developers. Continuous Deployments, Canary Releases, and other techniques allow for the fast deployment and rollback of software.

Developing iteratively is the de facto standard now, although the specific process you are using might vary. Which process you are following is secondary to how well it is executed and that it allows for continuous process improvement.

Do not focus on which process you follow, but on how fast you can iterate, learn, and adjust.

Chapter 148
Establish Change / Requirements / Configuration Management

If you want to build good software, you need to express what you want to build, you need to track it, and you need to manage it by versioning it.

Agile/Scrum focuses on user stories for expressing requirements; other methodologies go into various levels of depth and details using other formats; some industries might require specific technical specification formats.

Configuration management involves knowing the state of all artifacts that make up your system or project, managing the state of those artifacts, and releasing distinct versions of a system.

You need to know what you build, with what version of the code, on what version of the environment.

Many companies use popular source control systems, such as Git, and think they are home free. But many companies cannot tell you the difference between what you released today versus what you released last year, let alone extract a version of last year's release.

Change control and configuration management are more involved than just picking a source control system.

Chapter 149
Use of Component Architectures and Design Patterns

Depending on your technology stack, you must pick and choose the appropriate architecture for your application. However, over the last ten years or so, we have seen a consistent trend to use component technologies, web services, microservices, etc.

Things will keep evolving. For example, there has been a general move away from XML towards more lightweight JSON when it comes to REST APIs.

More and more we are seeing the consolidation around core technologies and approaches, such as component technologies, web- and microservices, as well as commonly used design patterns such as Twelve-Factor App. All provide best practice guidance for your applications. This is useful knowledge of what works, what does not work, and why.

Most importantly, these standards and patterns help you to not reinvent the wheel but to stand on the shoulders of others. This combined with the latest cloud technologies makes application development easier than ever before.

Chapter 150
Use Source Control

Disregarding the often emotionally charged discussion about tool preferences, let me state from my perspective that it does not matter if a team uses Git, SVN, CVS, Team Foundation Server, or any other source control system—as long as you use it effectively.

There is no doubt that, as of today, Git (and GitHub/Bit-Bucket) is the leading source control/version control system in the market. And for good reasons.

Having said that, all of the mentioned systems have advantages and disadvantages, but in the end, they are very much like choosing an Uber to get you to the airport—some are nice, some are big, some are small, but they all get you there.

It is not the specific tool you use that will make you successful, it is how you use it.

Chapter 151
IMPLEMENT UNIT TESTING

A best practice for constructing code includes building on demand (or at least the daily build) and smoke test.

Martin Fowler went further and proposes continuous integration that also integrates the concept of unit tests and self-testing code.

Kent Beck supposedly said that "daily builds are for wimps," meaning only continuous integration would provide the best benefits from a coding perspective.

Ultimately the frequency needs to be decided by the team, but the more frequent, the better. The reason for this is simple: It provides instantaneous feedback to the developer and allows for rapid problem resolution. Even though continuous integration and unit tests have gained popularity through XP, you can use these best practices on all types of projects.

Standard frameworks to automate builds and testing, such as Jenkins, Hudson, Ant, and xUnit, are freely available. Remember, unit tests are the foundation of all other kinds of testing.

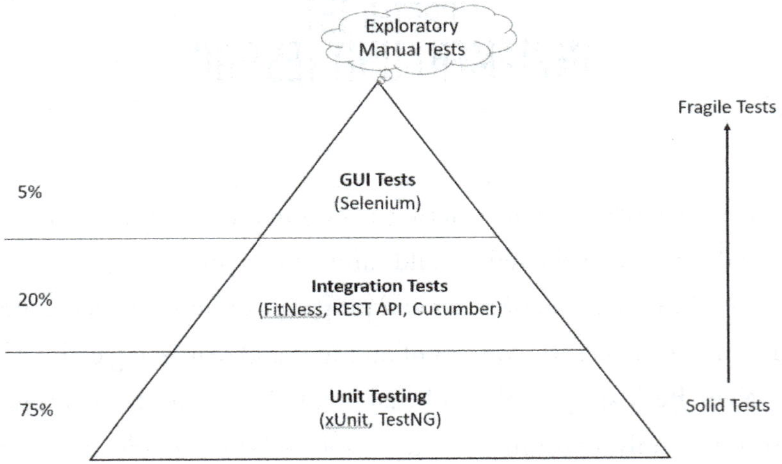

Figure 17 – Software Test Pyramid

Chapter 152
Know Your Code: Conduct Peer Reviews, Code Reviews, and Inspections

Make sure you know what is in your **source code**!

It is important to review other people's work. Experience has shown that problems are eliminated earlier this way and reviews are more effective than testing—and cheaper on a cost per defect basis.

Especially if your team works with contractors or a consulting service provider, review their work to guarantee knowledge transfer.

Reviews and inspections are also effective as a learning/teaching tool for your peers.

Cheap and effective, enabling knowledge transfer, and enabling learning/teaching—what else do you need to know?

Chapter 153
Perform Testing—Even in Agile/Scrum

Although Agile/Scrum proposes to have testing be the responsibility of the team and to make testing your User Stories part of the Definition of Done, there is a role for testing outside of the current sprint.

Disciplined testing in conjunction with extensive unit testing and reviews guarantees that software works. It is an integral part of software development that needs to be planned, executed, and whenever possible, automated.

Platform proliferation, however (various operating system versions, language versions, browsers, desktop vs. mobile platforms, etc.) has made testing critical to any product—and this kind of platform testing is frequently performed outside or after the current sprint. As are regression, security, and performance / scalability tests.

Although some companies have reached automation levels allowing them to be able to run all kinds of tests "in-sprint," many have not and are still following a test process that has tests run by a sprint + n offset.

Last but not least, testing still requires a good amount of manual, exploratory, effort.

Chapter 154
Automatically Measure the Process

How many developers are committing code every day? How many lines of code or class files? How many unit tests are executed per continuous integration run per day?

As the project is coded and tested, the defect arrival and fix rate can help measure the maturity of the code. It is important to use a defect tracking system that is linked to the source control management system and the automated build system.

By using defect tracking, it is possible to gauge when a project is ready to release by correlating information about the volatility of the codebase to the defects found/fixed based on the set of tests executed.

This kind of measurement has to occur automatically. Especially with continuous integration and automated platform tests, generating metrics needs to be automated due to the high frequency and large data volume that gets generated.

Chapter 155
Use Smart Tools

There are more and more effective tools that help software development teams address all kinds of problems, from coding standards to security scans. Use as many of the static and dynamic code analysis tools in as many automated ways as you can to make your life easier.

Chapter 156
Establish Test-Driven-Development

Test-Driven Development (TDD) is a software development process that relies on the repetition of a very short development cycle where requirements are turned into very specific test cases, then the software is improved to pass the new tests.

This is opposed to software development that allows software to be added that has not proven to meet requirements—many times introducing instabilities.

The key point here is that each requirement is developed based on test criteria and that you implement tests for every change.

It is hard to stress how much this approach improves quality, system stability, and maintainability—it is the basis of building a quality product from the ground up—especially for new projects.

But, what to do if the project has been active for eight years and has one million lines of code but no automated unit tests?

Do as the boy scouts do—whenever and wherever you touch code, go through the TDD approach and leave behind unit tests. Every time you add code, add unit tests. Every time you refactor a piece of code, create unit tests. That way you will slowly transform your codebase.

Chapter 157
Instrumentation

Instrumentation at all levels is critical for keeping a handle on your software, your infrastructure, and ultimately your customer satisfaction.

Every system is built differently, and every specific domain has its own challenges. However, at any given time you should be able to tell:

- How many users are on your system?
- What their activity level?
- What is the number of transactions that succeeded or failed?
- The utilization numbers for various server-side health indicators, like memory utilization, database connections, uptime, disk space utilization, etc.

Put this kind of instrumentation data on a big information radiator right next to your development team and you will be amazed what insights you will gain.

Chapter 158
Continuous Integration

Continuous Integration (CI) is a development practice that requires developers to integrate code into a shared repository several times a day.

Each check-in is then verified by an automated build, allowing teams to detect problems early, and fix them fast, shortening the feedback loop.

CI also allows for 3,000-mile-high monitoring of the development team—especially important in today's distributed development model—such as understanding who checks in code how frequently, who breaks the build, how many tests fail, how many tests are added, etc.

Development teams sometimes complain that they do not feel comfortable being monitored like that, but the reality is that in a highly distributed environment spread across several time zones or continents, this kind of data is crucial to gain visibility about how things are progressing.

Ultimately this kind of information is useful for teams to self-calibrate, cross-train, and take corrective actions when things go wrong. The result is a quality product.

Chapter 159
Automated Builds

Build automation is the process of automating the creation of a software build and the associated processes, including compiling computer source code into binary code, packaging binary code, and running automated tests for smoke testing.

Continuous Integration only tries to accomplish its tasks within the Continuous Integration environment-resolution of coding issues-such as defects, integration conflicts, merging and branching, unit test failures, etc.

Automated builds are usually a separate set of activities from Continuous Integration because automated builds focus on packaging the software solution into something consumable for a target environment (for example by automatically deploying binaries to a set of virtual machines or Docker containers for testing purposes).

Chapter 160
Automated Testing

No, **automated testing** does not mean that the software automatically tests itself!

Test automation requires tests to be coded by engineers so that tests can be automatically executed.

There are several implications with this: You are writing code, so you will need to apply the same software engineering concepts like design, refactoring, source control, etc. You will have to account for maintenance activities since test code, just like your other code, is rarely ever done. You will have to hire the right engineers to do this kind of coding. No, you cannot just forget about manual testing, because manual, exploratory, testing is still required, especially in the early days of development.

Automated testing tools and frameworks are capable of executing tests, reporting outcomes, and comparing results with earlier test runs. Tests carried out with these tools can be run repeatedly, at any time of day. This, in combination with CI/CD techniques, allows for 24/7 test cycles.

Be aware that although automated testing delivers great benefits to a software engineering effort, it does not come without a cost.

Chapter 161
CONTINUOUS TESTING

Continuous testing is the process of executing automated tests as part of the software delivery pipeline to obtain immediate feedback on the business risks associated with a software release candidate.

That is a mouthful. But what does it mean?

It means:

- You test all the time.
- You are using a Continuous Integration/Continuous Delivery pipeline to run automated tests.
- You have automated Unit Tests.
- You have automated Functional Tests.
- You have automated Integration Tests.
- You use dynamic and static code analysis tools.
- Test status is automatically reported at the appropriate level-unit, functional, integration.
- Exception reports are automatically created.
- Every time code changes, continuous testing kicks off.

Continuous testing assumes close to 100% automation. You

cannot do this manually. Why not? You want to cut down on the feedback loops, so you can immediately fix problems after they are introduced into your codebase. Manual test processes can easily take weeks, so automation is the only way to achieve this. If you do not automate, you cannot continuously test.

Chapter 162
Behavior Driven Development

Behavior Driven Development (**BDD**) is an agile software engineering practice that supports the discovery of and encourages collaboration about features to be implemented. How?

BDD relies on examples expressed in natural-language constructs expressing the expected system behavior. This is accomplished by using a standardized language called "Gherkin."

The advantage of Gherkin is that it uses natural language, i.e. English or one of thirty-seven other languages it has been translated to—as such, it is humanly definable and readable, with no programming knowledge required.

You might have come across Gherkin when you see system behavior expressed in a "Given… When… Then…" syntax.

This textural natural language description acts as a) documentation and b) input into the tools that generate skeleton automation code (i.e. Cucumber or SpecFlow). Once the automation code is refined by test engineers, it allows for automated validation of the system behavior as well as automated acceptance testing and regression testing. During your sprints, BDD enables the shifting of test activities to the left, which is core to achieving agility.

Chapter 163
Test-Driven Development is a Software Design Technique

Confusion abounds about **Test Driven Development (TDD)**. Depending on how technical people are, you might hear all kinds of crazy statements, like "TDD is all about the testers driving development to code better!" or "TDD is about creating unit tests." Wrong!

Test-driven development is a software design technique, not a test technique. It has nothing to do with software testers or testing.

When Kent Beck published his classic book, *Test-Driven Development by Example*,[1] back in 2002, his intention was to enable developers to write better code, to continually test and refactor it, as to avoid the old break/fix cycle that every engineering organization was stuck in.

TDD turns the development process on its head by implementing, in code, tests first and then implementing the code that is passing the tests. Once tests pass, as a programmer you know you have functionally working code.

With functionally working code, you can document it, refactor it to be more maintainable, and optimize it for performance. All those coding activities are validated by the original tests you implemented—so the programmer knows that things still work.

CHAPTER 164
CONTINUOUS DELIVERY

Continuous Delivery (CD) is an approach in which teams produce software in short cycles, sometimes as little as hours or minutes, ensuring that the software can be reliably released at any time—it is the effective combination of various techniques.

It aims at building, testing, and releasing software faster and more frequently. This assumes good software design using techniques like Test-Driven Development, a disciplined unit testing approach using tools like JUnit (or any other XUnit variant, depending on your implementation language), test automation at various levels using Behavior Driven Development, DevOps automation of the CI/CD pipeline, and use of many modern technologies like cloud computing, virtualization, Docker, Kubernetes, etc.

This approach helps reduce the cost, time, and risk of delivering changes by allowing for more incremental updates to applications in production. Even organizations that are not planning on continuously delivering software will greatly benefit from implementing these techniques.

Establishing the process and continuously exercising it makes perfect.

Chapter 165
Canary Release Rollouts

Canary releases are designed to reduce the risk of introducing a new software version in production by slowly rolling out the change to a small subset of users before making it available to everybody.

A good example of this is how Android app publishers can configure their app for Google Play Store Beta distribution, but the approach can be done regardless of the technology used.

I used this very approach 25 years ago when shipping an accounting software package to some 35,000 users–we would ship the new software (on CDs) to 2,000 users and wait for feedback. Then the next 3,000 copies would be sent out, waiting again, and then finally we released to the remaining user base.

Another approach is the Blue/Green/Deployment or A/B/Deployment, where you have two similar production environments, one with the current version (Blue/A), one with the future version (Green/B), and you switch users from one environment to the other without their knowledge.

Regardless, the purpose is to make the deployment of software releases transparent to the end users and minimize risk to the company.

Chapter 166
Use Docker Containers

In the old days, "Environmental Drift" was a big issue—basically the slow but unstoppable modification of operating systems due to patches and installed software. Every time software was installed, your development system and the target production system were drifting apart. Developing and testing on production-like systems, down to the specific operating system patch level and supported libraries was difficult to achieve and required god-like system administrative skills.

The running joke between developers was "It works on my machine!"

Virtual Machines allowed development to work on standardized images and made things easier and cheaper due to shared hardware.

With the introduction of Docker containers, we finally were able to encapsulate application-specific components and be guaranteed that they would run just like they did on the development system, regardless of where they were deployed to.

It is hard to overstate how many problems have been resolved by using Docker containers. If you are not using Docker containerization, you need to seriously investigate its use. Your life will never be the same.

Chapter 167
Use Automated Feedback Loops

Automated feedback loops are closely related to the Instrumentation Best Practice.

It is important to stress the need for automation here again—this cannot be a manual effort; it has to be automated. Too many times I have observed people scrambling to assemble PowerPoint presentations based on some cryptic information in log files—that does not work, it is too labor-intensive, error-prone, and welcomes inconsistencies due to transcription errors.

Feedback loops have to run constantly, update dashboards, and alert the right folks at critical junctures. For example, if a system runs "hot" and runs out of memory, an alert has to go out to your DevOps engineer on call to review what is going on; that DevOps engineer might, in turn, pull in developers to see what is wrong.

The automation might even take down the server and bring another server online. Unless this is automated, there is no way the feedback loop will be effective—it will be too slow, and in the above example, your hot running system will crash.

Chapter 168
User Stories: Focus on Personas and Avoid Poor Language Choices

Everybody knows the standard "As a <user>, I want to <action>, so that <benefit>" format. Simple enough. Then, why are so many User Stories just flat out useless?

Usually, it comes down to three areas of neglect:

1. People do not fully understand the INVEST and SPIDR principles.
2. The team did not think through User Personas. Nobody cares about generic users; you need to make User Stories. relevant by creating meaningful User Personas. Defining several personas that represent real users of your system will enable you to write better User Stories that have real meaning and provide better information for the implementation team.
3. Poor language choices. Excessive use of generic words that provide no specificity (car, plane, cloud, etc.), and frequent use of conjunctions and connectors (and, or, if, when, but, then, as-well-as, etc.)

When User Stories do not seem to convey the right

information, look at those three areas and the majority of your problems will resolve themselves. To find out more about writing good user stories quickly, check out my article "Quick & Dirty Guide to Writing Good User Stories" on LinkedIn.[1]

Part X
REFLECTIONS

The following reflections represent my personal musings about Agile, Scrum, agility, and industry trends as I see them.

As I pointed out in previous chapters, your opinion and viewpoints are shaped by your professional experiences. I happen to have entered my professional life on the tail-end of the mainframe area in the 1980s and have ever since ridden every technology wave that came along.

In hindsight what is important is to realize that technology waves are just that, waves. They come, and they go. Things keep changing, but things also stay the same. The latest thing will soon be the technology of the past. Some technologies change surprisingly little (relational databases), some revolutionize the world (iOS and Android, machine learning, self-driving cars) and become new standards.

Here are the things that are universally true:

- **Innovation acceleration**—Things change at an ever-faster pace. For example, twenty years ago you would have been considered crazy wanting to do image processing on millions of images – the processing power and algorithms just were not up to the job to do it in an economically viable way. Nowadays, you can process millions of images in minutes. Other examples include Tesla's rapid factory buildout and market penetrations, and the latest rapid COVID-19 vaccine development. Innovation is accelerating.

- **Interdisciplinary integration** – More and more disciplines are combined to create incredible value. For example, decoding the genetic code required biologists and computer scientists to work together. Your smartphone is a marvel of hardware engineering, software engineering, satellite communication, wireless communications, and probably five other disciplines that were all considered separate fields just twenty years ago. Political scientists now work with computer scientists and data scientists to analyze social media reactions. Electric cars combine more distinct technologies than I can count. Combining disciplines creates value.
- **Increasing reliance on black boxes** – More and more people use technology, and fewer and fewer people know how it works. This is a great benefit to users but also a great risk for developers. Nobody needs to know exactly how an Android phone works as long as it is useful to the user – it enables users to access a world of information, be better informed, be connected, document events in pictures and movies. That is a great benefit. However, in our world of increasing automation, more and more decisions are made based on "black boxes."

For example, try to understand why your credit score went up by twelve points or dropped by twenty-four; and good luck calling about that to get clarification. Or why does your GPS route you a certain way? More and more black boxes are used by us every day, and most of the time we do not know how they work, but we rely on the information provided by them.

- **Organizations are changing**—Companies have always organized certain ways to meet market demands—based on their understanding of management best practices at the time. From early management models by Max Weber, over scientific management introduced by Frederick Taylor, and Edwards Deming's Systems Theory, organizations are changing and adopting new ways of working. All these management theories have built upon each other, from the industrial revolution in the late 1800s to today's Systems Theory. However, despite all the advancements, the original top-down organization structure and Taylorism are the prevailing management structures that deeply affect how Agile and Scrum are adopted. Every time change is introduced, friction occurs. Agile or Digital transformations will expose organizational challenges.

I encourage everyone to reflect upon the following chapters in light of the above.

Chapter 169
The Agile Tide Is Rising

I have published books, articles, and countless posts about Agile/Scrum and am frequently asked why I am not more upbeat about the state of Agile.

The short answer is that I am very upbeat, just that when I try to teach people about Agile or bring a specific point across, frequently it is most effective to show what does not work well. We all seem to learn best from failures. But overall, the Agile world seems to be doing great.

Despite all the criticisms about Agile that are voiced on public forums like LinkedIn, despite all the infighting that happens between followers of one process framework over another, despite commercialization and watering down of Agile certifications, despite all kinds of other issues that are easily criticized, I can positively state this:

The big Agile tide that has risen over the last 20 years, and which is still rising, has lifted all the boats. We are undoubtedly in a better place today than we were in the late 1990s.

We all are standing on the shoulders of giants, and the fact that we all have vigorous discussions about agility, agile mindsets, and one specific technique or the other, is a sign of how far we have come. Let's celebrate it!

Chapter 170
Trust... Agile's Achilles Heel

Trust is Agile's Achilles Heel. No trust, no agility. Why?

The lack of trust is the ultimate CYA (Cover Your A..) because we want to make sure we are not to blame if a project goes off the rails.

Trust between people or between management and teams is usually built over time. But in the beginning, it requires a leap of faith, just like you allowing your teenager to drive off in your car the first time around.

I have never ever seen people come to work saying, "I want to do a horrible job today!" If you help them build rapport and trust, you will build a better team! It really is that simple.

No trust, no agility.

Chapter 171
Agility Is in the Eye of the Beholder

Agility is relative.

I worked on small, nimble, mobile phone apps where a single team had daily sprints and monthly releases to the app stores. I also worked in the automotive industry, with hundreds of teams, four-week sprints, and three-to-five-year release cycles. One of my clients produced gas turbines that had a product release cycle of seven years or more.

One size does not fit all.

Different companies have different ideas and views about what agility is.

Not all people have the same opinions as to what it means to be agile or attain agility.

The perception of agility is subjective—and mostly driven by our own professional experiences. The history of an industry frequently frames what is considered "normal," or "fast," or "slow." Industry disruptors frequently ignore that kind of framing or come with another perspective.

It is not possible to judge agility objectively.

What one person finds agile may not appeal to another—some might even call it reckless.

Ultimately, the market, your customers, and your competition drive your interpretation of agility and the resulting success—or failure.

Chapter 172
Agility's Past Is Prologue

In the 1990s I was unabashedly optimistic about Agility. Object-orientation had taken over as the dominant programming paradigm, the Waterfall process was generally acknowledged to be a dead end, and competing ideas were vigorously discussed in companies, publications, and industry conferences. The **Agile Manifesto** was consequently published in 2001.

But since, certification authorities and process frameworks have commercialized the original ideas and replaced the PMI Waterfall process with whatever Agile process. PMI acquired Disciplined Agile in 2019. The separation of technical Agility from business Agility has led to a split in focus.

So I am afraid that just as the original phrase was used in Shakespeare's *The Tempest* our past actions as Agile practitioners, proponents, and followers of one Agile framework or the other, one certification or the other, one technique or the other, will ultimately lead us down the path of killing Agility. Past is prologue.

Already many companies that adopted Agile or Scrum look suspiciously more like the old PMI Waterfall process. What have we gained? Where do we go from here?

Chapter 173
Agile Snake Oil: Beware the Agile Acronym Soup

Beware of agile acronym soup and their proponents that claim Scrum, Agile, SAFe, DAD, Nexus, Lean, Spotify, XP are all different methodologies or only theirs can solve your problems.

To make a long story short, Toyota invented the Toyota Production System, which formed the basis of what now is called "Lean." This established many of the Lean processes and terminology used today. W. Edwards Deming was instrumental in popularizing some of these approaches and systems thinking post-WWII.

Based on this and other work, Jeff Sutherland and Ken Schwaber created Scrum and published their work in the mid-1990s. Both later participated in a workshop that documented and published the **Agile Manifesto** in 2001. Sutherland published the Scrum Guide, and various other participants wrote books about Scrum or Scrum-related activities or disciplines.

Some of the methodology variants focused on scaling **Agile/Scrum** processes across many teams. It has been twenty years since the Agile Manifesto was published, and Agile processes frameworks/methodologies/techniques have multiplied like little rabbits ever since.

However, at the core, ***almost all use Scrum at the team level.***

There is lots of terminology noise, many opinionated proponents of one over the other - but very few practical differences.

Figure 18 – Agile Acronym Soup

Chapter 174
The Mindset: What Being Agile Really Means

Many companies now talk about "doing the **Agile** thing" or "going Agile" but what does that really mean? How come so many companies that have "gone Agile" still deliver the same products, at the same price points, at the same slow pace?

What if I told you that in 1989, I worked for a custom software house that followed a traditional waterfall development model but was able to deploy builds to their test environments on demand?

Our unit tests were fully automated, running with every build.

The build process was fully automated, allowing for dozens of system installations for every build.

System and Integration tests were fully automated.

Automated systems monitoring was in place.

We were Agile! In 1989!

Agile begins with attitude, not buzz words. You need relentless drive, execution focus, and a willingness to question tired old assumptions in order to improve your business. Or, as Peter Drucker said, "Culture eats strategy for breakfast."

Using a set of new ceremonies and buzz words will do very little to drive meaningful change.

Chapter 175
The Core Of Agile Is Always About Two Things

Cutting communication paths and delivering value faster.

Executives providing strategic direction and teams having the autonomy to implement it.

Transparency and empirical data.

Experimenting to find better ways of doing things and pivoting if the experiment fails.

Failing fast and recovering fast.

Recognizing people are the most important asset and paper plans are just promises.

Fewer meetings and more time to get work done.

Simplicity and purpose.

Chapter 176
"Hypocrisy Exists In The Space Between Language and Action."

Best. Quote. Ever!

Accountability sits at the center of Agility, and there is no better book to describe this than *Winning with Accountability: The Secret Language of High-Performing Organizations,* by Henry J. Evans.[1]

Chapter 177
The Big Agile Hustle...

We have hit "Peak Agile." Looking at the ever-increasing acronym soup and the "certification authorities" which multiply like bunny rabbits, I wonder where will it end?

Are we going to see 100+ Agile certifications? Seems like the only way to stand out of the crowd nowadays is not to have any certifications! Of course, that means your resume will not make it past the recruiting checkpoints. So, are Agile certifications just recruiting standards?

I have personally coached people that flat out told me that they were certified Scrum Masters or Product Owners but had no idea what to do. At least they asked. Have all those trainers that train Agile courses ever delivered anything in an Agile way?

I cannot help but feel that all this leads to a very prescriptive, process-focused, approach. Agile terminology abounds, embedded in waterfall thinking. We are handing out recipes instead of teaching the intent. We are talking about steps in a process, instead of the objectives we are trying to accomplish.

Seems like the Agile process has suddenly become more important than Agile thinking, attitude, and mindset. Or truly improving upon the existing process.

Chapter 178
Agile Zealots: Why Are So Many Agile Practitioners Not Agile?

"Can our daily standup go longer than fifteen minutes?"

"No."

"Can we have somebody on the team we dedicate as a tester?"

"No."

"Can we move the retrospective to the first day of the next sprint?"

"No."

"Can our Scrum Master cover two teams?"

"No."

"Can some of our team members only be 50% allocated?"

"No."

No, no, no, no, no!

Definition: Agile Zealot, a person who is fanatical and uncompromising in pursuit of their Agile ideals. You might as well call them Agile Wingnuts! Folks like this give Agile/Scrum a bad reputation.

I prefer Software Development Realpolitik, where decisions are made based on considerations of given circumstances and factors, rather than explicit Agile ideological notions.

Drucker supposedly said that "Culture Eats Strategy for Breakfast!" I think that realism and pragmatism eat Agile philosophy every day.

That is what Agile/Scrum are really about. Being Agile versus Doing Agile.

Beware of Agile Zealots professing to purity and righteousness of one recipe or the other. We cannot just follow a prescribed step-by-step process to success. Critical thinking is still a requirement.

Chapter 179
Understand If You Are in Your Opening-, Mid-, or Endgame

Everything has a beginning, middle, and end. You can think of this in terms of life, a soccer game, your car, or a myriad of other situations. In large Agile or Digital Transformations, the same applies.

For your opening game, you need to lay the groundwork, create buy-in, build consensus, and to some extent standardize some things so you can provide guidelines to all the teams. Your teams start laying out high-level Epics and Features based on prototypes and Spikes. Initiation.

In your midgame, you, your teams, and your stakeholders are finding their stride. You are fleshing out functionality and getting good at writing User Stories. Your teams start working well together. Your stakeholders are getting more comfortable. Stabilization.

In your endgame, you are polishing off the product. You are focused on making things work at scale, run real production data, and focus go-to-market activities. Finalization.

Whatever the specifics for your Agile or Digital Transformation, keep in mind where you are in the process so you can zero in on the relevant activities!

Chapter 180
TRUE IN ANY DOMAIN:
"Don't Ask The Barber Whether You Need A Haircut"

Warren Buffet has an uncanny ability to describe universal truths, and the above quotation is no exception. Beware whom you ask a question to—a barber will always tell you that you need a haircut!

Chapter 181
"You Can Observe a Lot by Just Watching"

This quote by Yogi Berra goes deeper than you might think. Who knew that Yogi Berra understood a thing or two about Lean Management?

You can discover the majority of issues in process problems or quality control by simply observing things.

In lean manufacturing, the idea of Gemba ("actual place" or "scene") is that the problems are visible, and the best improvement ideas will come from visiting the "actual place" or "scene." The Gemba Walk, much like Management By Walking Around, is an activity that takes management to the actual place of work and discover opportunities to improve upon the current process.[1]

Gemba walks denote the action of going to see the actual process, understand the work, ask questions, and learn. It is also known as one fundamental part of Lean management philosophy.

This does not just apply to work processes; you can figure out how well a team works together by having a meal and just observing their interactions—teams that laugh together, work well together.

Chapter 182
The Fallacy of Software Metrics

The problem with software development metrics is that most people think that metrics represent reality. They do not.

Same as "GDP Growth" the "Consumer Price Index" or "Income Inequality" tell you nothing about the real economy, software development metrics at best provide some directional information.

Metrics are an abstract representation of what we believe software development looks like. They are part of a model we believe represents the reality—but they are not the reality themselves.

This is the same concept as Alfred Korzybski's "The map is not the territory" or Alan Watts's "The menu is not the meal."

Something to keep in mind next time we review our software development metrics and jump to conclusions.

Chapter 183
Understanding Black Swan Events

People are really bad at anticipating the unforeseen.

We think we can manage risks by creating simplistic risk lists and risk prioritization schemes, but the reality is that we simply pretend in order to feel good.

Understand black swan events by reading *The Black Swan: Second Edition: The Impact of the Highly Improbable* by Nassim Nicholas Taleb.[1]

Read at your own peril if you love math, statistics, and philosophy. Prepare for your mind to be blown and old assumptions to be torn down.

Chapter 184
How to Be Successful with Agile / Scrum.... Grit

Intuitively we all know that success to a large part is determined by grit.

All things being equal, you will encounter smart people across the board, gifted people, super athletes, super communicators, etc. And what you will find is that success will not automatically come to the obviously smart, gifted, athletic, or best communicators but to the ones that have grit.

Agile success is very much the same way. If you do not succeed at first, try again, and again. Companies, teams, and people that have grit will succeed with Agile/Scrum. It is a matter of committing to and sticking with it. Angela Duckworth's Ted Talk "Grit: The Power of Passion and Perseverance" provides valuable insight into why this works.[1]

Chapter 185
Product Roadmaps: Vision Without Actionable Plan Is Fantasy

Product Roadmaps must translate into Epics which in turn get decomposed into User Stories. Continuous refinement of User Stories into digestible, actionable, pieces of work are at the core of Agile and Scrum.

Beware that the focus on little bite-sized work items can cause the old "can't see the forest for the trees" problem. This is normal as team members operate at the implementation level and are worried about producing working software.

Product Owners however need to make sure that higher-level Epics represent a Product Roadmap that you are trying to implement.

Depending on what agile tool you use and how you structured your work, you might have additional hierarchical groupings. For example, some tools let your group items according to the SAFe terminology: Epic – Feature - Story. Think of it simply as a functional decomposition of work.

User Stories represent the actionable plan because you can deliver them in successive sprints. Without refined User Stories, you do not have an actionable plan. Without good Epics that layout the high-level work, you do not have a roadmap.

Chapter 186
Product Owners: Saying No Is Good

Inexperienced **Product Owners** frequently take customer or stakeholder requests and literally translate them into epics and user stories. You know, the customer is always right! Wrong.

Product Owners' real job is to triangulate between what customers want, what the team can deliver while being profitable. If profitability or product ROI are not of concern, the Product Owner still needs to worry about working within budget constraints.

Many years ago, I worked on a system that had many, many, reporting requirements. We received hundreds of requests for reports. After doing the back-of-the-napkin calculation, we figured out it would take us several years to deliver all requested reports. The solution was to provide a report writing interface to our customers, so they could generate the reports themselves—not to create hundreds of reports.

What is the moral of the story here? Customers most of the time do not know what they want, so Product Owners need to say "No" a lot. But in turn, they come up with innovative solutions to satisfy the customers' needs.

Chapter 187
Sometimes Things Do Not End Well

Sometimes you do everything right from an Agile and Scrum perspective, but things still do not work out.

History is littered with top of the line companies that followed Agile and Scrum but got killed in the market—conversely, sometimes companies follow horrible processes but are widely successful.

The lesson is simple: Sometimes things do not end well. Sometimes the market overtakes you. Sometimes your competitors are just flat out better, faster, or more innovative. Sometimes you start focusing more on your process rather than serving your customers. There are probably as many reasons why companies fail as there are failed companies.

Agile or Scrum helps companies manage their work better, creating better, more optimal outcomes. Agile or Scrum provides a set of techniques that make work results more predictable. But, following Agile or Scrum is no guarantee for success.

Chapter 188
You Are Responsible for Your Feelings and Actions

Stephen R. Covey said that between stimulus and response there is space. In that space is our power to choose our response. In our response lies our growth and our freedom.

Hypocrisy exists in the space between language and action. If you do what you say, your influence will grow. If you do not do what you say, hypocrisy will reign. Henry J. Evans helps you understand how to harness the power of your words in his book *Winning with Accountability: The Secret Language of High-Performing Organizations* 6th Edition.[1]

Chapter 189
Build Trust and Do Not Burn Your Bridges

Trust is built over time. Conduct yourself with integrity and honesty. Praise folks that work with you or for you. Take responsibility for failures. Engage openly. Do not speak behind people's back. Speak plainly. Be nice.

If you do these things, people will want to work with you. Again, and again.

If you do not, at best folks will not remember much about you. At worst, they will not want to work with you again.

Chapter 190
Beware of Ancient Management Techniques

According to www.gantt.com, "The first Gantt chart was devised in the mid-1890s by Karol Adamiecki, a Polish engineer who ran a steelworks in southern Poland and had become interested in management ideas and techniques.

"Some fifteen years after Adamiecki, Henry Gantt, an American engineer and project management consultant, devised his own version of the chart and it was this that became widely known and popular in western countries. Consequently, it was Henry Gantt whose name was to become associated with charts of this type."[1]

Do we really believe that using a technique invented in 1890 is still relevant in 2020?

Chapter 191
You Cannot Make Yearly Plans Any Longer

Old way of doing things: Annual Roadmaps—the organization's portfolio plan, as well as the release plan or product roadmap, are created once a year in advance. This is so 1890s. It assumes incredibly slow-moving markets, slow-moving competitors, slow-moving everything.

New way of doing things: Continuous Roadmap Refinement—Agile planning is always "continuous." At the portfolio level, the plan needs to be revised at least every three months, ideally every six weeks.

But what about budgets? How can you re-budget so frequently?

You do not. You set the budget once a year, and then you deliver against the budget by flexing around scope, always delivering the highest priority items first. Because things are moving so fast, you deliver high-value items, and low-value items either fall off the radar or are not needed any longer because of changing market conditions.

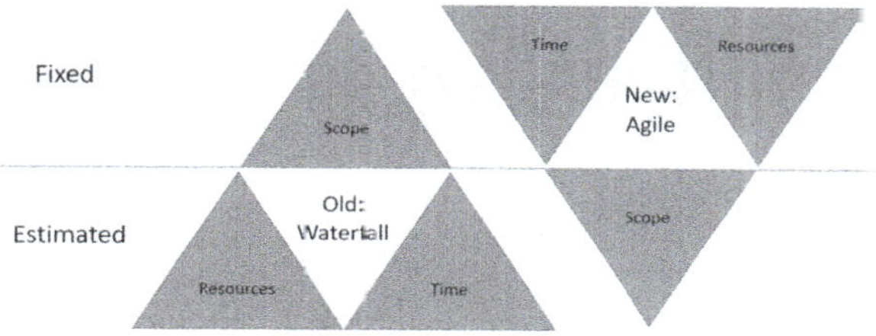

Figure 19 – Annual Roadmaps versus Continuous Roadmaps Refinement

Chapter 192
It's a Framework, Not the Bible!

Whenever folks that discuss Agile and Scrum come together, whether, in person or online, you will hear discussions like this:

"SAFe clearly states that you need to …." or "Where in the Scrum Guide does it say….?" or "You cannot do Scrum without a Scrum Master!" or an infinite amount of other variations and claims.

All of this is made worse by a plethora of Agile/Scrum certification authorities that teach and certify, for $300 to $5,000, any number of folks willing to sit through various levels of training classes and pass a middle school level exam to get a digital badge.

The reason it makes it worse is because it has factionalized the Agile community. You have the SAFe© faction, the Scrum@Scale© faction, the PMI DAD faction, and probably another 10 smaller variants (Crystal, LeSS, etc.) as well as the "Independents" (like me!) who claim you do not need to buy into any of the big frameworks but can make up your own mind.

Long story short: Everybody has their favored framework and although all of them have the same underlying principles, they are not the Bible! You can pick and choose to fit your environment.

Chapter 193
Agile Frameworks: They Are Not Recipes!

SAFe, Scrum, Crystal, Disciplined Agile, and other popular agile frameworks are often viewed as step-by-step descriptions for success. If you follow the ceremonies and rules step-by-step, you are successful. Wrong!

Most are simply frameworks, with suggested approaches, ceremonies, rules, etc. You own your success by implementing some or all the suggested approaches. No framework can do that for you.

Think of agile frameworks as cooking techniques. You broil, grill, sauté, steam, smoke, slow cook, or pressure cook your food. You apply a technique to food, following a recipe. Good cooks learn techniques first, then look for recipes—and modify recipes to their own taste.

If business success was as simple as following a recipe, what would we need smart people for? Look at your business, look at your challenges, and then use the best agile techniques to be successful.

Use the frameworks as a guide for your company's recipe for success. Do not search for an elusive recipe that nobody outside of your organization can create. Own your success by implementing your unique process with proven techniques

Chapter 194
Agile Paradox: Companies That Need Help Most Will Benefit Least

Severely dysfunctional companies usually do not benefit from an Agile or Digital Transformation.

This is a paradox that parallels human patients. If you are severely sick, an exercise program will not help you. You need to have a certain base level of health and physical capabilities to benefit. For example, you need to be able to walk to improve and gain the next level of fitness. If you are so sick that you are bedridden, you need emergency procedures.

For companies, the equivalent of emergency procedures might be the replacement of executive officers, the sale and break-up of the company, or bankruptcy proceedings.

If your company suffers from mismanagement, unethical business practices, employee intimidation, poor or unfair hiring practices, eroding sales and profitability numbers, or a myriad of other issues, no Agile or Digital Transformation effort will help you.

Relatively well run and healthy companies on the other hand will be able to build upon their success and make substantial improvements. Companies that need help most will benefit least.

Chapter 195
Agile Paradox: People Have Contradicting Needs

The human aspects of an Agile or Digital Transformations are fascinating. Anybody that has gone through a substantial change, whether in their personal or professional life, has experienced the contradicting needs all humans have.

We all want a certain amount of certainty and comfort. We want to know that our job will still be there tomorrow and that the job has not changed. Most of us enjoy some amount of uncertainty and variety in our lives, so we go on road trips or hiking tours in the Andes.

We strive for significance and want to be recognized by our peers. All of us want to be loved and feel connected.

Everybody wants to grow—who wants to get stuck with their teenage knowledge and understanding of the world? And finally, we all somehow want to contribute, and that comes in many forms.

Therein lie the challenges for many Agile or Digital Transformations—they severely impact these desires, upending careers, changing daily routines, reshaping peer groups, and often making us question our value.

Ignore these human needs and desires, and you run the risk of sabotaging your Agile or Digital Transformation.

CHAPTER 196
"PEOPLE DO NOT RISE TO THE OCCASION, THEY SINK TO THEIR LEVEL OF TRAINING"

Attributed to an anonymous Navy Seal and the Greek lyrical poet Archilochus, this quote highlights the inability of people to operate at a reliable high level of competence unless they obsessively prepare and train for it.

When we face stress or difficulties, we all revert to our levels of comfort and familiarity. The most common examples include the alcoholic ordering a drink, the ex-smoker lighting up a cigarette, a dieter starting to binge eat, the athlete suddenly losing technique, a ball room dancer losing her rhythm, etc.

Stress is the great equalizer. This is totally normal and equally applies to organizations. Organizations that have been undergoing Agile or Digital Transformations frequently revert to their old way of doing business as soon as stress creates pressures.

The trick is that you need to "level up" and create a new baseline, so that when stresses cause you to revert, you fall back to a higher level of performance than what you started out with. Once you have a new baseline, you go on to improve on it again. And again.

Chapter 197
Remote Work Impacts Agility

Agile and **Scrum** are built upon some basic underlying principles: small teams, local decision making, people over process, face to face communication, etc.

Problem is that just as Agile and Scrum came about, just as the **Agile Manifesto**[1] was published, the world encountered an unprecedented combination of international labor arbitrage (offshoring to countries with cheaper labor costs) and vastly improved communication and networking capabilities. Hence, remote work and remote teams were born. For knowledge work, this internationalization and offshoring of work has continued at an accelerating pace.

Agile and Scrum practitioners are faced with people and/or teams working overseas, sometimes in extremely distributed teams. The recent COVID-19 pandemic has only accelerated this process with more people working from home.

What does this mean? Everybody acknowledges that face to face interactions are the preferred way of communicating. Remote work is not as productive. But there is no "going back" this is the new normal. For many reasons, not just COVID-19.

Chapter 198
No More Secret Roadmaps

In the old days, companies safeguarded their roadmaps both internally and externally, to ensure that confidential information would not leak to the market or competitors. It was not uncommon to have roadmaps made available to staff just once a year, providing only limited visibility as to what had to get implemented over longer-term time horizons.

Although confidentiality is still required, sharing longer-term plans and roadmaps with staff is essential to enable productive and smooth operations on a long-lasting product or value stream. You can no longer keep your roadmaps secret as the continuous planning process requires visibility.

If the portfolio planning and the release plan or the product roadmap are not visible to everybody, you cannot predict where you are going—coordination problems abound. This old-style "information hoarding" is counterproductive.

Having visibility to and understanding the roadmap is crucial for any Agile team, and it needs to be available to everybody at any time.

Chapter 199
User Stories Are Like Pictures

User Stories are generally acknowledged to be the lowest or most granular level of functional decomposition you can achieve in a Backlog.

The purpose of a User Story is multi-faceted:

1. It is supposed to describe from a business perspective what needs to be implemented.
2. It is supposed to act as a conversation piece for team members to discuss what is involved in the implementation.
3. Based on the conversation, the team is supposed to confirm its understanding of what needs to be done.

Card – Conversation – Confirmation. User Stories are not supposed to be detailed requirements documents—think of User Stories as pictures.

A picture by itself is neutral unless you took the picture yourself; a photographer or painter remembers all kinds of additional information, like the location, the weather, and the mood of the subject. That is what the conversation piece gives team members—the conversation fills in the details necessary for the implementation.

CONCLUSION

Agile and **Scrum** are simple—you can compare Agile and Scrum to the game of golf—very few rules are written, but it is hard to play well. This is dramatically different from Baseball, which has many rules but is equally hard to play well.

My point is that the lack of clearly defined, prescriptive, rules does not make it easier to play the game. In many ways, it makes it harder.

I come back to the beginning of this small book. You need to **understand the basic ideas** that are underlying the Agile and Scrum movement, and really understand them, which is not hard because they are just a few.

Then you must **contrast those basic ideas to the unlimited number of myths that are being propagated**, not because they are wrong but because many parties have their own interests in mind.

Some only do Scrum, so they only argue for Scrum. Some are dedicated to SAFe, so you will only hear arguments supporting SAFe. Some are training organizations that make money training and certifying Scrum Masters, so their interest is not in making Agile and Scrum successful but in getting as many people as possible certified. Some are staff augmentation companies that only care about placing contract workers. Others are top-flight consulting organizations that use Agile and Scrum as "differentiating" service offerings to close multi-million-dollar consulting engagements. I could go on.

If you understand the myths and why third parties are interested in propagating them, you will be better prepared to be successful with Agile and Scrum—for your particular situation. Be clear-eyed and realistic about things, do not buy into the myths. Changing the way you work is hard, so the better you prepare, the better your chances for success. There is no simple, off-the-shelf solution you can buy and be successful with.

Lastly, **you need to clearly understand the facts and principles to make sense of things**. The **best practices mentioned simply provide you with a list of generally acknowledged practices that modern software development organizations follow**. You do not need to follow all of them to deliver good software, but if you follow only a few, you might not realize the benefits of Agile and Scrum.

I could have easily structured this book in a traditional way. Just like the **Agile Manifesto**,[1] I could have laid out principles, followed by a conclusion.

Instead, I focused on the myths and facts before principles. Why? In this new world of "alternative facts" and opinions, I wanted to address the big disagreements and variations that you frequently come across when entering the Agile tent. It can be confusing.

I believe it's important to understand the principles so that you know what to shoot for. They are goals. We strive to achieve them.

Myths and facts, on the other hand, do not tell us what we can achieve. They restrict us. They limit us and show us the challenges we might face.

I am a practitioner, so I wanted to give you something tangible and real. Not something theoretical. Many a project has died an untimely death because there were high principles with no realistic and tangible outcomes.

As I pointed out at the beginning, understanding the core underlying ideas of Agile and Scrum will enable you to make sense of all of this. Again, they are:

1. Optimized Communication.
2. Small Batches of Work (Small User Stories).
3. Ownership and Personal Accountability.
4. Small Teams Equal Better and Faster Communication.
5. Localized Decision Making.
6. Continuous Learning.
7. Value Streams and Wait Times.
8. A New Way of Thinking, Planning, and Budgeting.

These eight ideas drive everything in Agile and Scrum.

1. Why optimized communication? Because small teams, working closely together, with only a few short ceremonies, focusing on working software will communicate progress more effectively than the best status report could.
2. Why have small batches of work? Small user stories? Because it has been shown that small batches of work create more effective flow and better throughput, leading to better results.
3. Why have ownership and personal accountability? Because it creates better products, faster, and makes people have skin in the game. It also makes for happier employees and all the benefits associated with that.
4. Why have small teams? Because they communicate more effectively. The larger the team, the less communication is going on.
5. Why work locally? Because localized decision making is faster. As soon as you have to reach out to another city, state, country, time zone, all communication slows down, becomes outdated, and causes delays.
6. Why continually refine and improve on processes? Because only continual learning enables you to improve. The Plan – Do – Check – Act cycle of continuous improvement.

7. Why understand your value streams and wait times? So you can go from quarterly builds and yearly releases to releasing on-demand, by cutting, eliminating, and automating the wait times in your value streams.
8. How do you know that you have enough budget? You do not. But you do know that prioritization of your user stories in your backlog will force high-value items to the top of the backlog and drive low-value items down. Consequently, you will have delivered the best possible value at whatever budget amount you end up with

Look at each of these eight ideas when you are challenged, and I promise you will find the answer you are looking for.

I sincerely hope that this book has helped you understand Agile and Scrum better. Enjoy the journey, and most importantly, do not forget to have fun!

Notes

Chapter 1 – Optimized Communication

1. http://www.agilemanifesto.org

Chapter 6 – Continuous Love

1. https://en.wikipedia.org/wiki/PDCA Index
2. https://en.wikipedia.org/wiki/OODA_loop

Chapter 10 – Agility Guarantees Business Success

1. https://youtu.be/bNpx7gpSqbY

Chapter 17 – Do or Do Not. There Is No Try.

1. https://youtu.be/Ri-M_Vo3w5A

Chapter 19 – Quality Is Free

1. https://www.amazon.com/Quality-Free-Certain-Becomes-Business/dp/0070145121/
2. https://www.linkedin.com/pulse/delivering-good-product-increments-brian-will/

Chapter 23 – Management Is an Artificial Construct

1. https://en.wikipedia.org/wiki/Max_Weber

Chapter 29 – Good People Management Still Required

1. http://www.agilemanifesto.org

Chapter 61 – Simplicity

1. http://www.agilemanifesto.org

Chapter 64 – Lack of Executive Change

1. https://smallwarsjournal.com/jrnl/art/how-powerpoint-stifles-understanding-creativity-and-innovation-within-your-organization

2. https://www.theguardian.com/commentisfree/2015/sep/23/powerpoint-thought-students-bullet-points-information

Chapter 79 – It's the Manager

1. https://www.amazon.com/Its-Manager-managers-organizations-long-term/dp/1595622241/

Chapter 105 – The Agile Rashomon Effect

1. https://en.wikipedia.org/wiki/Rashomon_effect

Chapter 115 – Not Having a Definition of Ready (Or Definition Of Done)

1. https://www.linkedin.com/pulse/definition-ready-dor-vs-done-dod-brian-will/

Chapter 116 – The Product Owner Proxy or Product Owner Agent

1. https://www.linkedin.com/pulse/product-owners-what-makes-good-one-brian-will/

Chapter 117 – No Slack Time

1. https://www.amazon.com/dp/0767907698/

Chapter 127 – Forgetting About or Not Understanding INVEST

1. https://xp123.com/articles/invest-in-good-stories-and-smart-tasks/

Part X – Agile Best Practices

1. https://en.wikipedia.org/wiki/Best_practice

2. https://www.merriam-webster.com/dictionary/best%20practice

Chapter 141 – Use Risk-Based Test Prioritization

1. https://en.wikipedia.org/wiki/Pareto_analysis

Chapter 163 – Test-Driven Development Is a Software Design Technique

1. https://www.amazon.com/Test-Driven-Development-Kent-Beck/dp/0321146530/

Chapter 168 – User Stories: Focus on Personas and Avoid Poor Language Choices

1 https://www.linkedin.com/pulse/quick-dirty-guide-writing-good-user-stories-brian-will/

Chapter 176 – Hypocrisy Exists in the Space between Language and Action

1. https://www.amazon.com/Winning-Accountability-Language-High-Performing-Organizations/dp/0981924204/

Chapter 181 – "You Can Observe a Lot by Just Watching"

1. https://en.wikipedia.org/wiki/Gemba

Chapter 183 –Understanding Black Swan Events

1. https://www.amazon.com/Black-Swan-Improbable-Robustness-Fragility/dp/081297381X/

Chapter 184 – How to Be Successful With Agile/Scrum...Grit

1. https://www.ted.com/talks/angela_lee_duckworth_grit_the_power_of_passion_and_perseverance

Chapter 188 – You Are Responsible For Your Feelings and Actions

1. https://www.amazon.com/Winning-Accountability-Language-High-Performing-Organizations/dp/0981924204/

Chapter 190 – Beware of Ancient Management Techniques

1. www.gantt.com

Chapter 197 – Remote Work Impacts Agility

1. http://www.agilemanifesto.org

Conclusion

1. http://www.agilemanifesto.org

ABOUT THE AUTHOR

Brian is passionately committed to helping companies make the transition to Agile/Scrum and increased team productivity.

His most recent professional focus has been primarily on software development consulting within the Predictive Analytics/Machine Learning / Big Data space, Mobile App Development, and Enterprise Software.

In the past, Brian worked within the commercial software (COTS) product space in various functional capacities—covering diverse problem areas, from operating system and compiler development to enterprise financial systems and mobile apps. He also has substantial experience with Ecommerce systems transacting in excess of $3B.

Brian's background is rooted in the operating system and the object-oriented technology community. He started as a software engineer on UNIX and C/C++ architectures, led quality assurance teams for UNIX kernel development efforts, and managed several test engineering organizations for commercial Smalltalk compiler teams.

Based on his early exposure to Agile thinking within the Smalltalk community, Brian was an early adopter of Agile practices, including the early application of Kent Beck's now-famous xUnit framework for Smalltalk. As early as 1996 he was an implementor and adopter of DevOps practices.

During the course of his career, Brian has held many roles, including successful stints as a Scrum Product Owner, Scrum Master, Agile Coach, and member of development teams.

In addition, he has held numerous functional management roles: VP of Professional Services, Senior Program Manager, Director of Quality Assurance, and Director of Quality Engineering. He was also part of an early adopter team implementing real-time, in memory, predictive analytics for credit and debit card fraud detection.

His multifaceted background gives Brian the ability to understand (and explain) Agile/Scrum and its implications equally well from multiple perspectives: from the development team to the executive board.

Index